How We Behave

Graham Carter

Education Officer
Beaulieu Manor, Beaulieu

Richard Cox

Head of Science Department
South Kilburn High School

Ginn and Company Limited, Aylesbury

 047510
ISBN 0 602 21909 4 Product No. 518581713

Published by Ginn and Company Ltd,
Elsinore House, Buckingham Street,
Aylesbury, Bucks HP20 2NQ

Original research and design for this book by
Alphabet and Image, Sherborne, Dorset
Drawings by Elizabeth Winson

Phototypeset by
Tradespools Limited, Frome, Somerset

Printed in the German Democratic Republic by
Interdruck Leipzig

Contents

A thick layer of fur-like feathers helps the penguin to survive in an Antarctic environment which would kill most other animals.

Introduction

This is a book about survival—about staying alive. In the past few years, you have probably learned a lot about the structure of animals and about how their body systems work—the skeleton, the skin, the functioning of the digestive system and so on. You may also have learned something about the habitat of various animals and about how their body form is adapted to their way of life. However, you have probably found out very little about their behaviour, and yet the pattern of behaviour of any animal is the key to its survival.

All animals are active for a great part of their lives, and very little of this activity is wasted. It all helps, in some way, to keep that individual alive. Shrews and elephants both have to feed for most of their waking hours to avoid starvation. Swallows migrate to warmer climates in winter. Mice and voles must be constantly alert to the arrival of a hawk or a stoat. Similarly, a primitive fisherman stands by the water with his spear raised, and he knows that he must aim his spear under where the fish appears to be. At the other extreme, a housewife may simply walk to the nearest fish and chip shop to buy a meal for herself and her family. All these patterns of behaviour, and many more, help to ensure the survival of the individual.

In the pages that follow, we will be considering various aspects of animal behaviour in some detail. We will see how the individual animal receives information from its surroundings and how it reacts to the information received, and we will investigate how the individual animal reacts to other members of the same

species. We are animals, and throughout the book you will find references to human behaviour.

Do not expect always to find exact statements of fact, for the study of animal behaviour is, as yet, very far from an exact science, with a basic set of rules which can be committed to memory. Experimental investigation of behaviour can be very difficult, for, in most cases, behaviour in captivity differs markedly from behaviour in the wild state, and this more natural behaviour may be difficult to observe. The problems facing students of human behaviour are even greater. At the simplest level, most people object to being stared at (why do they, we wonder)—and their behaviour changes. At a more advanced level, problems arise because experimentation is more difficult. For instance, there would be a public outcry if a student of human behaviour began collecting human babies and rearing them in soundproof boxes without human contact. Nor can we draw exact parallels between the behaviour patterns of man and other animals, although this might be very tempting. The behaviour of any animal is affected by its body structure, its physiology, its intelligence and its surroundings. No other animal resembles man in all these ways. However, it is possible that studies of animal behaviour can give some insight into human behaviour by suggesting worthwhile areas for investigation.

Behaviour studies help to build up a balanced knowledge of the animal world and the same argument is equally valid when applied to studies in human biology. Man's behaviour should help him to survive but today it appears to be threatening his survival. It is therefore very important to try and understand why we behave as we do. It is hoped that this book will give you some insight into your own behaviour and also help you to sympathize with other people's behaviour. A sympathetic and knowledgeable outlook is essential if we are to deal with the social problems, such as mental illness, which confront us at this time.

The South American Indian's way of life provides everything that is necessary for survival. He is threatened with extinction by the encroachment of 'civilization'.

1 The Senses

To survive, an animal must be aware of its *environment*, or surroundings. The environment contains things the animal wants such as food, water, and warmth. It also contains many dangers which it must avoid. An animal uses its *senses* to find out about these things.

For example, when you cross a road you use your eyes to look at the traffic. In other words, you use your sense of sight to tell you about your environment. You obviously need the information to survive. How does the information about the traffic reach your brain?

Light is a *stimulus*, and it reflects from the cars and travels through the air to your eyes. Your eyes contain cells called *receptors* which are sensitive to the light. Each eye is linked to the brain by a nerve. The receptors respond to the light by producing *nerve impulses* which travel along the nerves to the brain. In the brain, these impulses cause the sensations of sight, and you see the cars.

Notice that the information about the cars reaches your brain in two stages: first, the stimulus carries the information from the cars to the receptors. Second, nerve impulses carry the information from the receptors to the brain. Later in the book you can read more about nerve impulses, but for the present it may help you to think of them as *information messages*. Let us look at each of the senses and see how they work.

A fly sees through thousands of tiny lenses on the surface of its eyes and gets a general picture like a mosaic. Although such compound eyes are unable to form precise images, they are extremely efficient at detecting movements.

Seeing

Animals need light in order to see. No animal, not even a cat, can see in total darkness. There are many ways in which animals use the information they receive from light.

When a fish-louse has a light shone on it from below it turns and swims upside down. Like many animals, it uses the direction of light to *orientate* itself, or fix its body position in its environment.

Fly-maggots have no eyes, but they are sensitive to light all the same. They hatch from eggs which the female fly lays in such places as manure heaps and animal corpses. The maggots must remain in these places where there is food, moisture, and shelter. Do you think fly-maggots move towards or away from light?

Ragworms are well known to fishermen who use them for bait. Ragworms have four eyes but they cannot see things as clearly as we can. However, they are able to see other animals moving, and this ability is useful in detecting mates, food, and enemies.

As you know, the human eye can form very clear images. Our eyes provide us with a great deal of information about our environment. Let us see how they work.

When we look at something, many rays of light travel from it, enter our eyes, and finally strike the *retina*. Look at the diagram which shows the parts of the human eye and try to decide which parts must be transparent if light is to reach the retina.

The retina consists of more than 100 million light-sensitive cells called *rods* and *cones*. These cells contain chemicals called *visual pigments* and, when light strikes

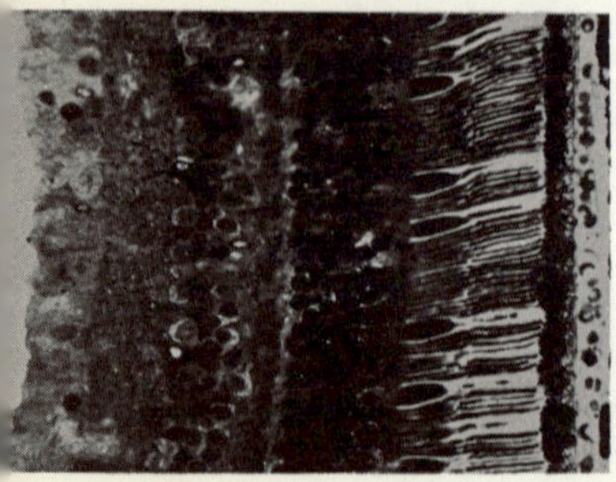

Light has to pass through several layers of retinal cells (above) before it reaches the rods and cones (thin and oval cells at the right). Alongside: section through the human eye.

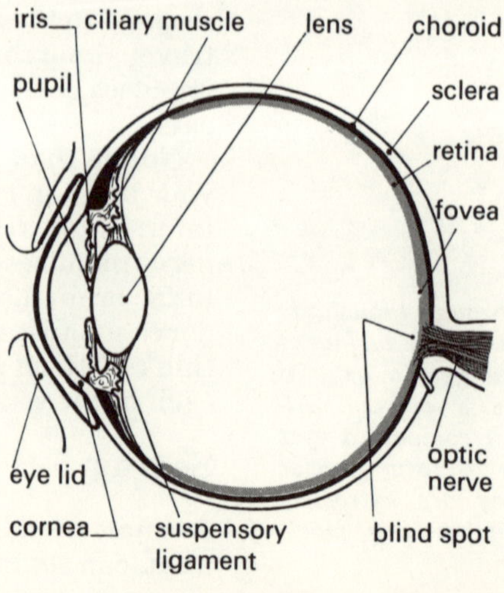

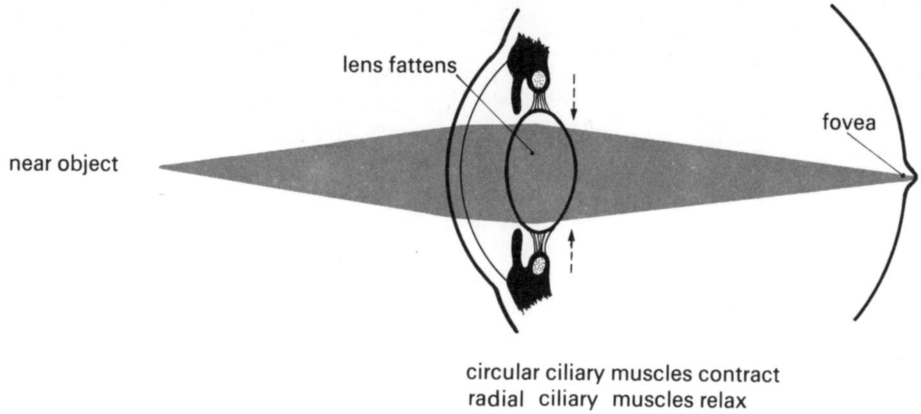

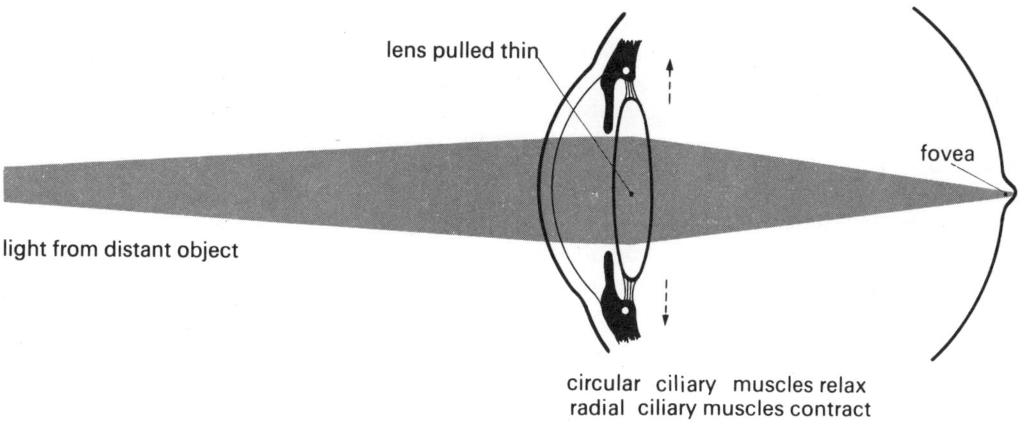

The shape of the lens is changed by the action of the ciliary muscles during accommodation.

the rods and cones, the pigments change into slightly different forms. These chemical changes cause impulses to travel along the optic nerves to the brain.

The light which enters the eye is *focussed* so that a clear image is formed on the retina. The light is focussed mostly by the *cornea* (the transparent covering at the front of the eye), but fine focussing is done by the lens. The eye cannot focus on close and distant objects at the same time. To focus on a close object the lens must be quite fat, but to focus on a distant object the lens must be made thinner. The shape of the lens is altered by the *ciliary muscles* and the process is called *accommodation*. Some people are not able to focus their eyes very well and they must wear glasses (lenses). Not all animals focus in the same way as we do. For instance, fishes focus by altering the distance between the lens and the retina.

Look at your eyes in a mirror and notice the black circle (the *pupil*) at the centre of the coloured circle

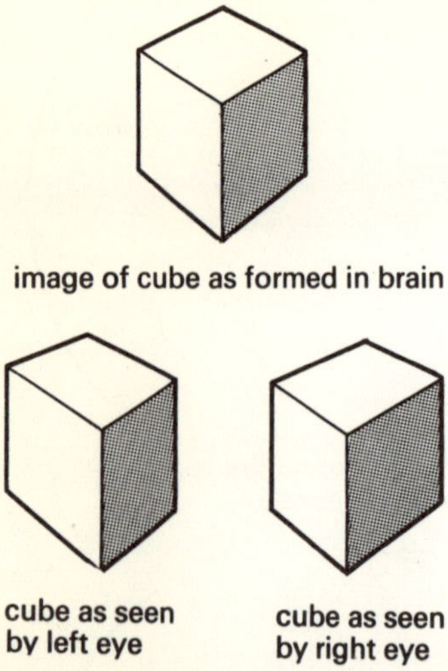

Above: the brain combines the images seen by each eye in a single picture. Below: the position of the rabbit's eyes give it a wider field of vision than that of man. How much do you think the frog can see when almost completely under water?

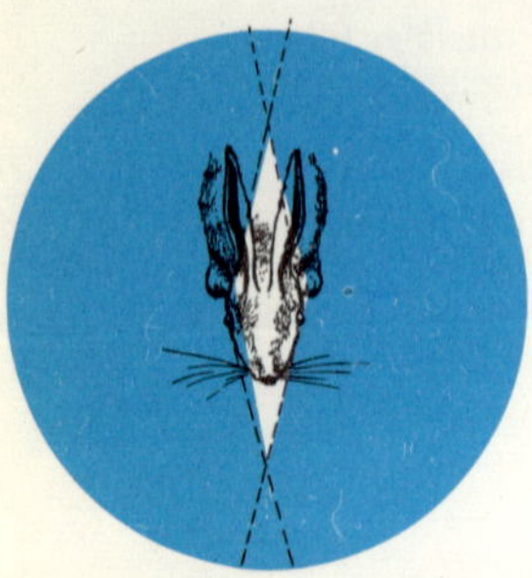

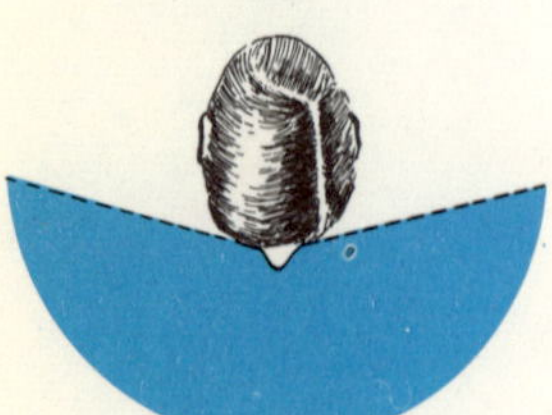

(the *iris*). The pupil is actually a hole in the iris and it allows light to enter the main part of the eye. The iris is able to move and so alter the size of the pupil. The pupil is made smaller in bright light, but in dim light it is made larger so that more of the available light can enter the eye. Animals which are *nocturnal* (active at night) usually have a slit-shaped pupil which narrows tightly in bright light to prevent the animal being dazzled.

Some animals (including ourselves) have both eyes at the front of the head. When both eyes look at something, each eye sees a slightly different view of it. The brain puts both views together to form one picture, which is three dimensional (solid in appearance). As both eyes view the object, they are turned inwards by muscles which send impulses to the brain. The brain can use this information to calculate the distance of the object since the closer the object the more the eyes must be turned inwards. This ability is very important for hunting animals which need to judge the distance of their prey before they spring on it.

Animals such as rabbits and most birds have their eyes at the sides of the head. Thus, they see things on either side of them with one eye only. However, with this arrangement, an animal can see objects all round it—even things behind it. This is very useful to preyed-upon plant-eaters because they can see dangers in any direction.

Let us return for a while to the retina of the eye. You may remember that it consists of rods and cones which respond to light by sending impulses to the brain. There are two other important facts about rods and cones. First, cones respond only to bright light whereas rods respond to dim light. Second, only the cones can 'tell' the brain what colours we are seeing. These two facts explain why we can see colours only by day—when the cones are working. At night we see everything in shades of grey (except bright lights) because only the rods are working.

Some people are totally colour-blind, which means that they have no sensations of colour. Some other people have partly defective colour-vision and cannot tell some colours from others. For example, some people cannot tell the difference between yellow and orange; others cannot tell blue from red. Many kinds of animals cannot see colours. Scientists have performed experiments which show, for example, that cats and dogs cannot see colours but that ants and minnows

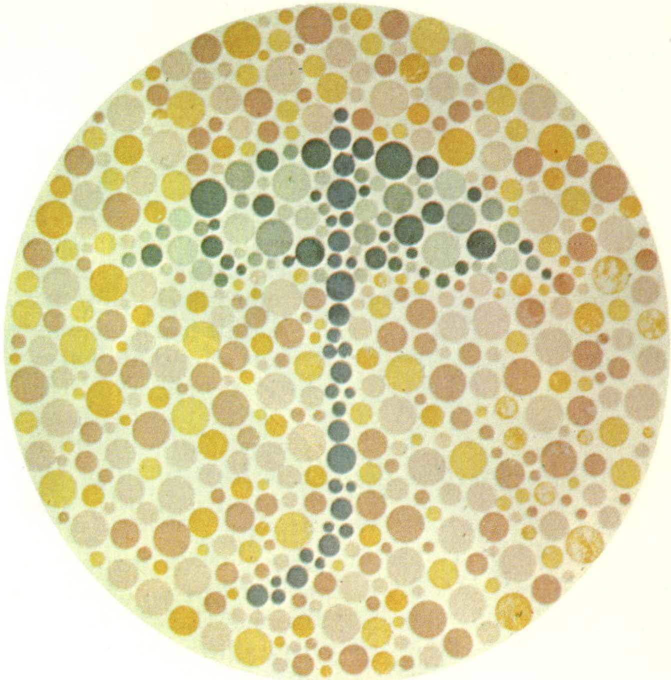

These pictures are used in conjunction with others in colour vision tests. Instead of seeing an umbrella in the top picture, people with colour defects see a hockey stick. Those with red-defective vision see only a spade in the lower picture, those with green-defective vision see only a fork.

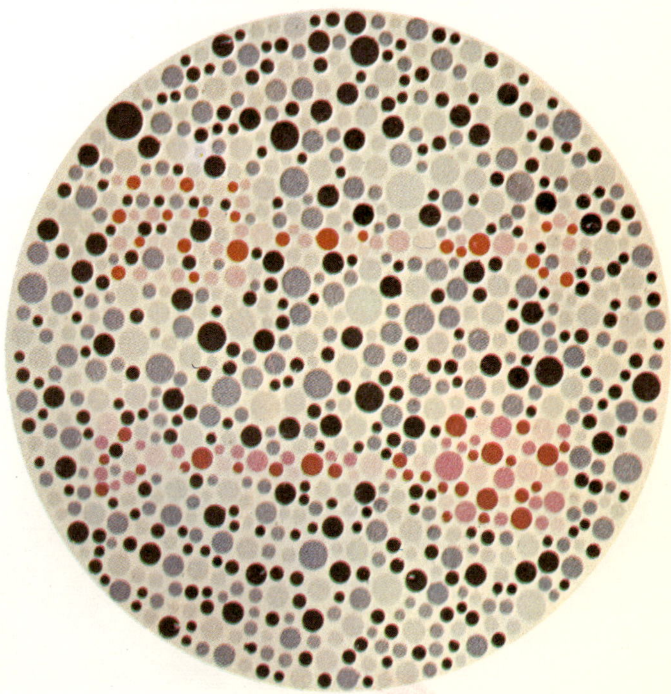

can. The ability to see colours can be very useful. Thus, insects use colour to recognize flowers which provide them with food. A female stickleback recognizes the male by the red colour of its belly.

Hearing

Eyes are not very useful in some situations. For example, you cannot see someone lurking behind a rock or moving about in pitch darkness. In these circumstances your ears are more useful.

At this moment you are using your eyes to read these words and to find out what we want to tell you. However, people also *communicate* by speaking and by listening to what others say. Therefore, hearing is a very important factor in communication. But what *is* hearing?

When you hit a table with your hand you cause vibrations in the table—these vibrations can be felt by someone who is touching the table. The table will cause vibrations in the air next to it, and the air-vibrations (sound waves) will spread to your ears. The ears contain receptors which are sensitive to the stimulus of sound. Each ear responds to the sound by producing impulses which travel along a nerve to the brain. Thus, your hand is sensitive to the vibrations in the table and your ear is sensitive to the vibrations in the air.

Some animals cannot hear. For example, the earthworm is not sensitive to sound waves—it has no ears. However, the worm can feel vibrations of the ground

Reptiles such as this lizard do not have pinnae to protect the ear and gather sound. The ear drum is in the dark hollow behind the eye.

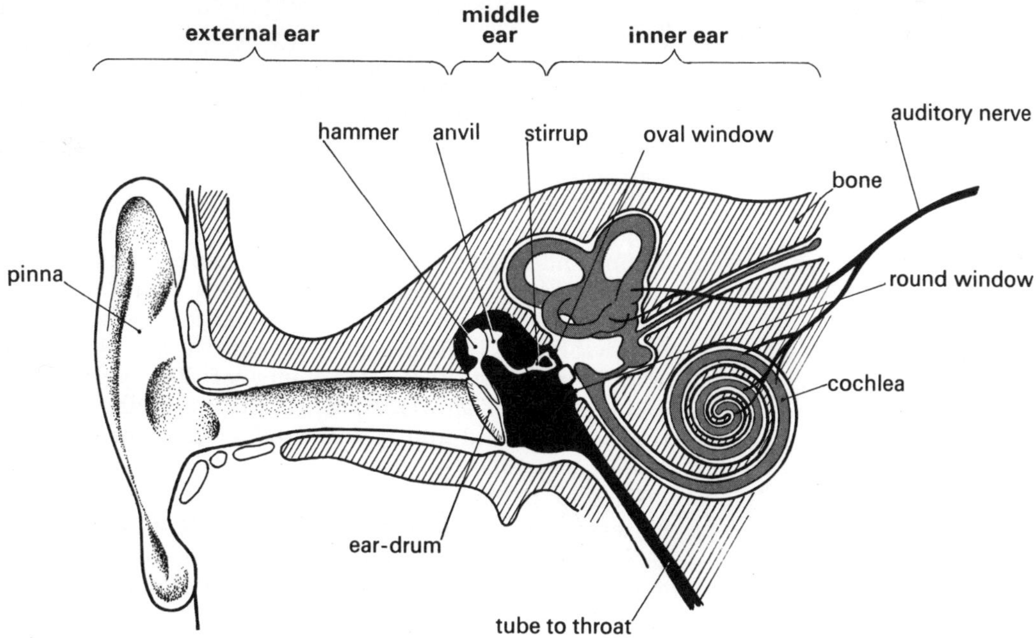

which warn it that another animal, such as a bird, is approaching.

Let us now turn to the human ear. The diagram shows the three sections of the ear—the external ear, the middle ear, and the inner ear. Notice that most of the ear is hidden inside the skull, where it is protected from damage.

When sound waves strike the *pinna* (ear trumpet), they bounce off it and into the air-passage. The passage contains hairs and wax to keep small insects out, and at its inner end there is a very thin cone-shaped skin called the *ear-drum*. When the sound waves hit the ear-drum they make it vibrate.

The ear-drum separates the outer ear from the middle ear. The middle ear is a space filled with air and surrounded by skull bone. In the bone on the inner side of the middle ear is a hole with a *membrane* (thin skin) stretched across it. This is the *oval window* and it connects the middle ear with the inner ear. The ear-drum and oval window are connected by three small bones called the *hammer*, *anvil*, and *stirrup*. When the ear-drum vibrates, the vibrations are carried through these bones to the oval window which also begins to vibrate.

The inner ear contains a structure called the *cochlea* which looks like a coiled shell. The cochlea contains fluid and receptor cells which are sensitive to vibrations. When the oval window vibrates it makes the

Vibrations from the ear-drum are magnified in the middle ear and then passed by the liquid in the cochlea to the sense cells, where they are converted to impulses which are carried to the brain by the auditory nerve. The old-fashioned ear trumpet, below, acted as an extension of the pinna, collecting more sound waves.

fluid in the cochlea vibrate, and the receptors respond by producing nerve impulses. The impulses travel along the nerve to the brain where they cause the sensation of sound.

Scientists believe that different areas of the cochlea respond to sounds of different *frequencies*. Perhaps you do not understand what is meant by the frequency of a sound? When you strike a tuning-fork it vibrates. In other words, it moves backwards and forwards. If the fork vibrates 300 times in a second we say that the sound produced has a frequency of 300 cycles per second (cps for short). Another fork may vibrate 400 times per second and it will produce a sound with a frequency of 400 cps.

Human beings can hear sounds with frequencies between 16 cps and 20 000 cps. You may have heard of a *Galton whistle* which is used to call dogs. This whistle can produce a sound with a frequency of about 25 000 cps. Dogs can hear this but we cannot. What is the point of a Galton whistle?

Bats listen in to their own high-frequency sounds to catch insects and to avoid obstacles in the dark. When they fly the bats produce some noises which we hear as squeaks and other noises with frequencies up to 100 000

cps which, obviously, we cannot hear. The high-frequency sounds reflect from objects to the bat's ears so that the bat hears echoes of its own sounds. The bat uses these echoes to locate the positions of insects and obstacles.

An animal may hear the sounds made by its prey, mate, young, and enemies. Next, it needs to find out the *source* (where the sound is coming from). How do animals do this? Some animals, such as horses can move their *pinnae* (ear-trumpets) to discover the source of a sound. Human beings cannot use this method because the necessary muscles are not well developed. However, you may know someone who can waggle his ears slightly!

We are able to locate sounds in the following way. A sound coming from, say, the left will reach the left ear slightly before it reaches the right ear. Also, the right ear will be shaded from the sound by the head and so the left ear will hear the sound more loudly than the right ear. The brain can use the information from the two ears to work out that the sound is coming from the left. Do you think this method works when the sound source is directly in front of or behind the head? In these cases, what can we do to locate the source?

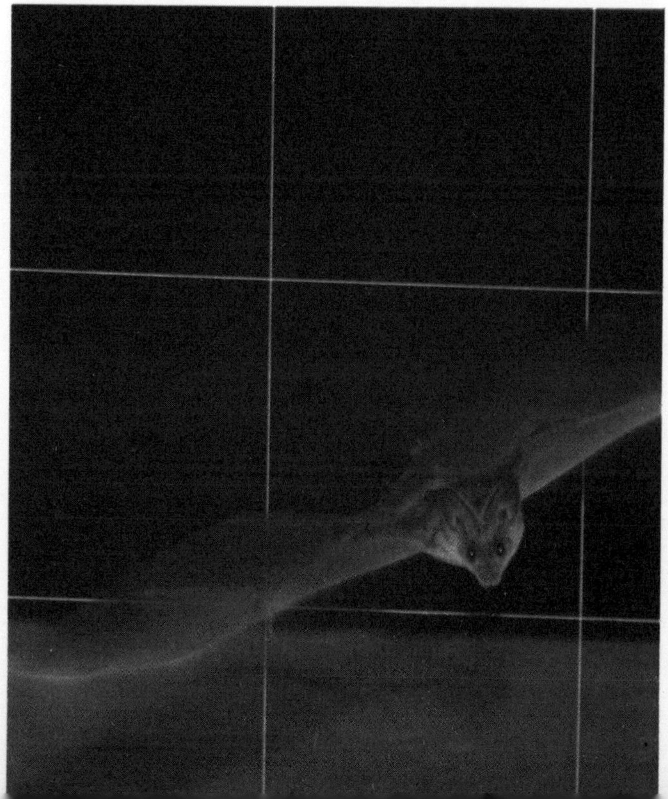

The bat is producing high-frequency squeaks which are reflected by obstacles such as the wire grid in its path and heard as echoes by the large ears. It is thus able to navigate in total darkness through a space considerably narrower than its wingspan. This true sequence was illuminated by electronic flash.

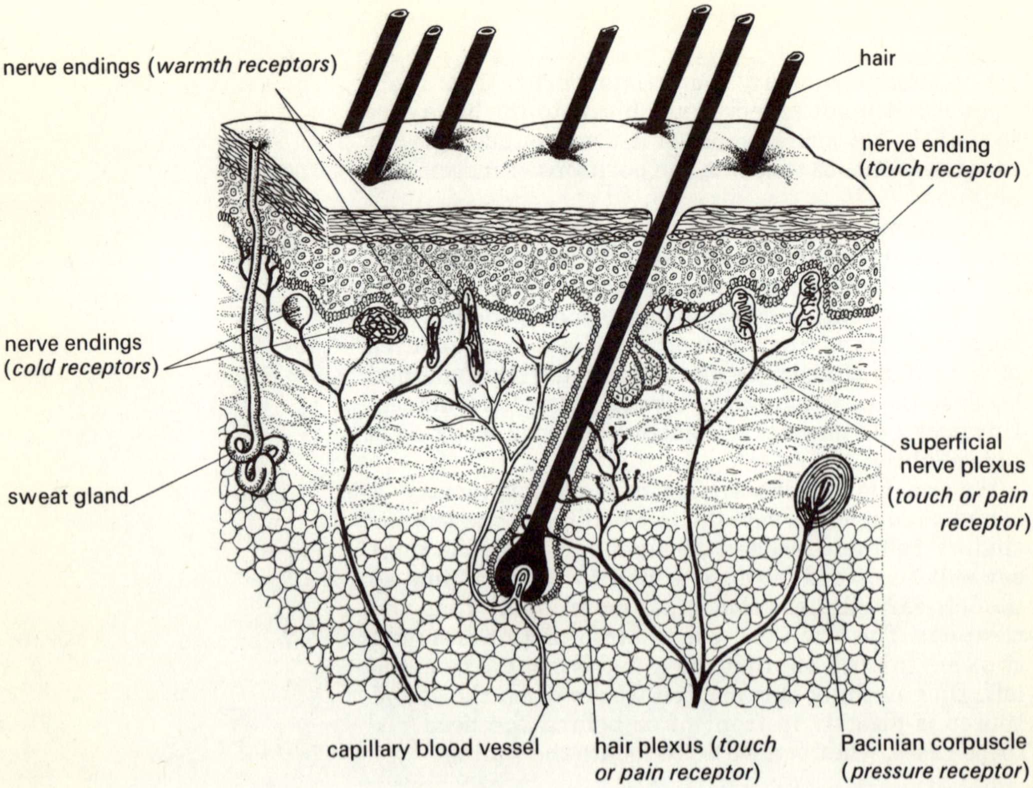

Section through human skin.

Touching

Our sense of touch gives us information about the objects with which we come into contact. We may feel an object lightly touching us or pressing hard against us. The object may feel warm or cold and it may even cause us pain. Perhaps we should not talk about *one* sense of touch because there are so many different sensations that we can feel. Why do we feel so many sensations?

In the diagram of a section through the skin you can see that there are several types of nerve-endings or *receptors*. These are able to send impulses along nerves to the brain. For example, when something presses hard against the skin, the *Pacinian corpuscles* send impulses to the brain. In the brain, these impulses cause the sensation of *pressure*. Notice that there are different nerve-endings for warmth and cold. When the temperature of the skin rises, the warmth nerve-endings send impulses. When the skin temperature falls, the cold nerve-endings send impulses. If the object touching the skin is very hot, the pain nerve-endings will also send impulses. All the nerve-endings have something in common; they are all stimulated by something which moves the skin in some way. For example, if you press your hand you can see the skin move. You cannot see the movements caused by a hot object because they are very small and take place within the skin.

Let us turn to a more pleasant sensation—the tickle! If you gently touch the hairs on the back of your hand or arm you will experience a tickling sensation when a hair is moved. The nerve-endings wrapped round the roots of the hairs send impulses to the brain where they cause the tickling sensation.

Other mammals have hairs supplied with nerve-endings. For example, on a cat's face there are long whiskers which are sensitive to touch. These whiskers are used to detect objects in the dark.

Our fingers are very well supplied with touch nerve-endings and so they are used to explore the surfaces of objects. A snake uses its flickering tongue for the same purpose. Whatever part of the body is used, the sense of touch is a very important aid to survival. When an animal is unexpectedly touched it usually moves away from the possible danger. If your hand touches something hot, you move it away so that it will not be damaged. You have only to sit on a drawing pin to appreciate fully the protective function of your sense of touch!

Although he cannot see, this boy has probably developed a better sense of touch and hearing than yours.

Tasting

When we eat food we taste it and smell it. The combinations of tastes and smells give food its various flavours. You will have noticed that food seems tasteless when you have a cold. The reason is you cannot smell the food and so it loses most of its flavour. Actually, your tongue can detect only four tastes: sweet, salt, sour, and bitter. The diagram shows the areas of the tongue which are particularly sensitive to each of these four types of taste, and may help to explain why you can taste a sweet substance better at the front than at the back of your tongue.

Your tongue is covered with small ridges called *papillae*. The sense organs of taste, called *taste buds*, lie in the sides of the papillae, and each taste bud consist of several taste cells (receptors) linked by nerves to the brain.

When you place food on your tongue, some of the food enters the grooves between the papillae and

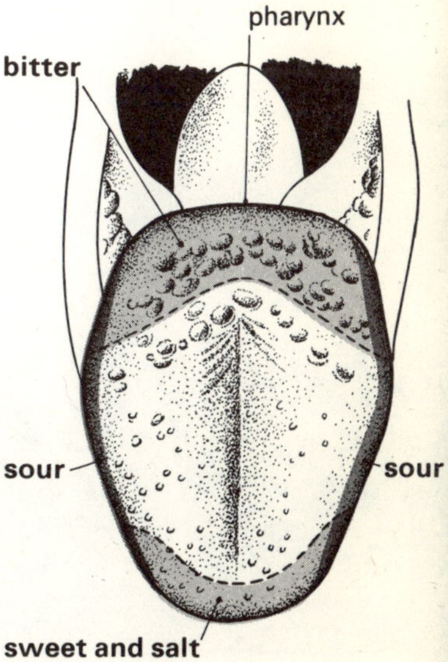

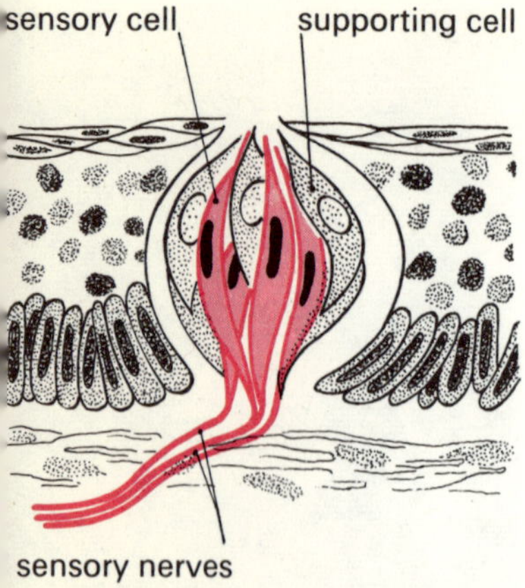

A taste bud on the tongue.

A police-trained labrador uses his nose to smell out drugs hidden in a chair.

comes into contact with the taste buds. The chemicals in the food act as a stimulus and cause the taste cells to send impulses to the brain, where they cause the sensation of taste. Your tongue also contains nerve-endings which are sensitive to touch, heat, cold, and pain. These enable you to experience the coldness of an ice-cream and the roughness of a crisp, but remember that these are really feeling, and not tasting, sensations.

Other mammals taste with their tongues as we do. However, fishes have taste buds not only in the mouth, but also on their gills, skin, and barbels. In fact, fishes can taste with their whole body surface.

Smelling

If you have ever smelled a bad egg then you have smelled hydrogen sulphide, a gas produced by rotting eggs. You have almost certainly smelled the vapour which is given off when petrol evaporates. Gases and vapours like these drift through the air and are drawn up the nostrils when we breathe in. In the upper part of each nostril there is a small area of skin, about a centimetre across, which contains smell receptors. These receptors are cells sensitive to chemicals carried in the air. Chemicals such as hydrogen sulphide and petrol vapour cause the receptors to send impulses to the brain. In the brain, the impulses cause the sensations of smell. As you know, when we want to smell something carefully, we sniff the air. Sniffing draws the air up into the part of the nostril where the smell receptors are found.

Our sense of smell is more acute than our sense of taste even though they both rely on the stimulus of chemicals. The human nose is so sensitive that you would be able to smell one hundredth of a milligram of methyl mercaptan (the smell of garlic) if it were released into your classroom.

A dog is even more sensitive to smells and is able to recognize one smell among many. The police make use of this and train their dogs to smell out hidden substances such as cannabis. Anyone who has taken a dog for a walk knows that dogs mark out their territory by urinating at frequent intervals. They also spend a great deal of time smelling the urine left by intruders into their territory.

A sense of smell is much more important than the sense of taste. It can be used to find food, mates and enemies when they cannot be seen. Many animals,

The sense cells in the antennae of some moths can detect minute quantities of chemicals carried in the air over long distances.

such as foxes, use their sense of smell when hunting their prey. Man's sense of smell is not good enough for this purpose so we use dogs to find prey for us. Do you know why experienced hunters stalk with the wind in their faces?

Many insects are attracted by the scents of flowers which provide their food, and the gypsy moth uses its sense of smell to find its mate. The antennae of male gypsy moths can detect the smell of a female gypsy moth 3 km away (Chapter 4). Ants recognize other members of their nest by their distinctive smell. An ant from another nest will have a different smell and will be attacked.

Other senses

The senses which you have been reading about provide us with much information of the world around us. We have other senses which give us information about what our own bodies are doing.

Close your eyes and try to touch the end of your nose with your right forefinger. Most people can do this, and it shows that we know the position of our arm and fingers without using our eyes. How is this possible?

The muscles which move our limbs are attached to the bones by connective strips called *tendons*. Where the bones meet there are joints which allow movement. In the muscles are *stretch receptors* which send nerve impulses to the brain when the muscles are stretched during movement. The tendons contain receptors which work in a similar way. There are receptors also in the joints, and these provide information about how the joint is being moved. Your brain then uses the

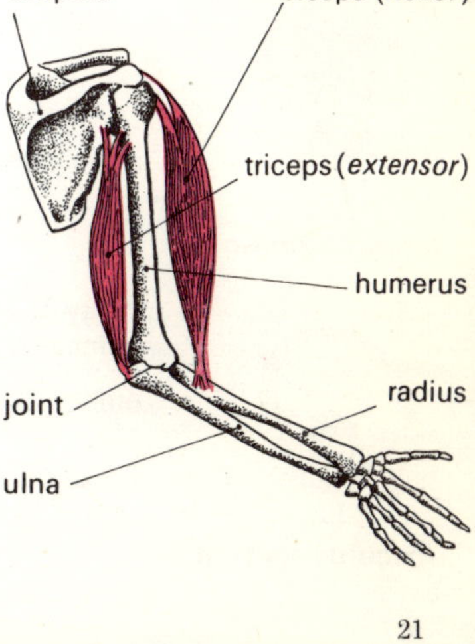

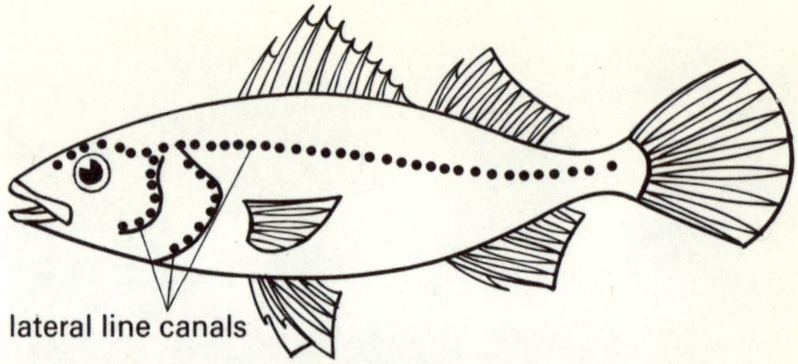

Sensory cells in canals along its body give a fish information about its position in the water and whether anything else is swimming near it.

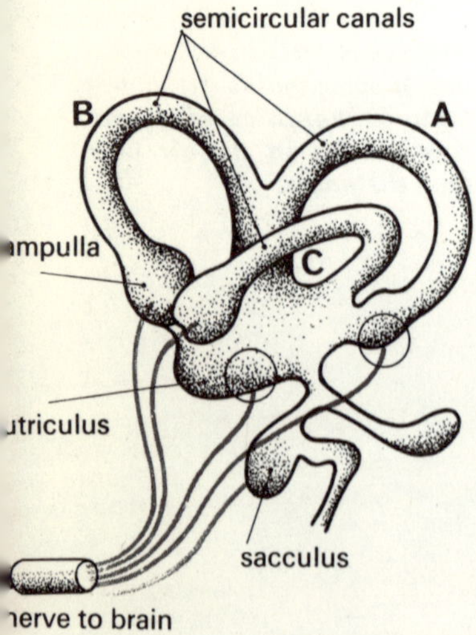

Impulses from the labyrinth tell the brain if the head is moving, and in what direction.

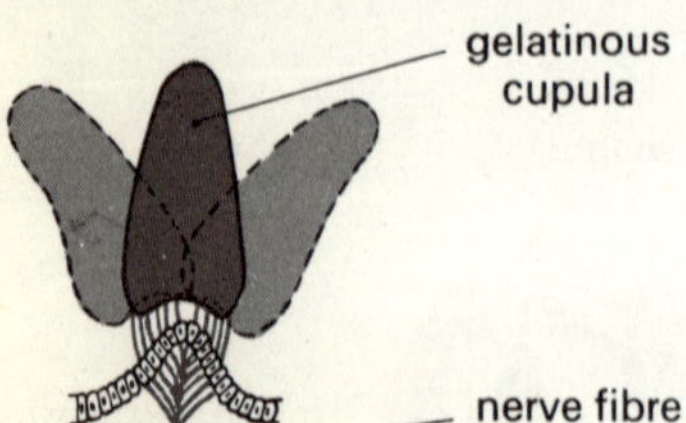

ampulla section

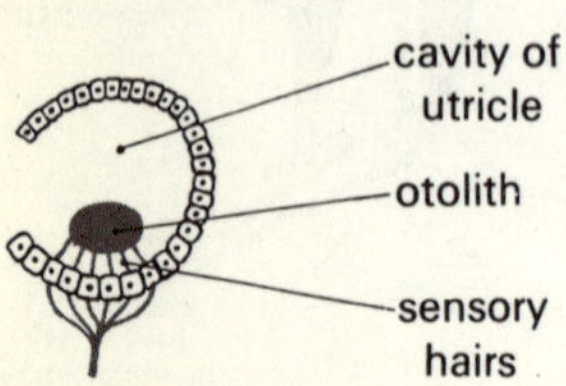

utriculus section

information from muscles, tendons and joints to work out the positions of various parts of your body.

Information about movements of your head is provided by a part of the inner ear called the *labyrinth*. The diagram shows the labyrinth and you can see that it is a complicated structure. Two parts of it are called the *sacculus* (little sack) and *utriculus* (little bag). Each of these contains patches of receptor cells. These cells have hairs which are attached to a small earstone (called an *otolith*) which is made of chalky material. When your head moves, the hair cells move. The otolith acts like a pendulum and moves against the hair cells. As a result, the hairs bend and this, in turn, causes the hair-cells to send impulses to the brain. The impulses give information to the brain that your head is no longer upright.

The picture of the labyrinth also shows three *semicircular canals*. At one end of each canal is a swelling (called an *ampulla*) which contains hair-cells. The canals and ampullae also contain liquid. When your head moves, the wall of the canal and the hair cells move, but the fluid lags behind slightly as it is not fixed to the walls—rather like when you rotate a cup of coffee. The fluid causes a drag on the *cupula*. This bends the hairs, causing the cells to send impulses to the brain, telling it how quickly the head is moving.

But how does the brain know the direction of the head movements? The three canals are arranged in the three planes of space in which the head can move. When your head tilts clockwise or anticlockwise, the hairs in canal A are bent. When your head tips in a forward-backward direction the hairs in canal B are bent. Finally, when your head moves in a left–right direction the hairs in canal C are bent. Since the brain 'knows' which canal is sending impulses, it is able to work out the direction of the head movement.

Thus, the labyrinth sends much information to the brain. It tells the brain when the head is no longer upright. It also tells the brain about the direction and speed of head movements. This information is important because we can keep our balance only if we are aware of what our bodies are doing. When the inner ear is completely destroyed by disease, the affected person will be unable to move about or stand upright in the dark. He will have lost his sense of balance because his brain will not be receiving information about the movements of his head. The situation is not quite so serious in daylight because his eyes will provide information about the position of his body.

When you stop spinning you become dizzy because the liquid in the labyrinth continues to move for a short time. The skater avoids dizziness by throwing her head back.

2 Information Received

In the first chapter, you read that an animal uses its senses to obtain information about its environment. The information may cause the animal to do something —in other words, to make a *response*. For example, the response of a rabbit on seeing a fox is to run away.

You may remember that we called light a stimulus because our eyes respond to it. We can also use the word stimulus to describe something which causes an animal to respond in a more obvious way. Thus, a fox is a stimulus which causes a rabbit to run away.

Not all responses are caused by stimuli from the environment around us. Some responses are caused by stimuli from inside the body. For example, when you decide to go for a walk the activity of your brain is the stimulus which causes your response.

The example of the fox and the rabbit shows how information from the environment can be used to make responses which are essential to survival. Let us now look at some of the different types of responses.

Inherited (unlearned) responses

The first type of inherited response is a *reflex action.* If you unexpectedly touch a very hot object with your hand you experience a reflex action. You withdraw your hand very quickly and *automatically*—in other words, you do not *decide* to carry out this response. Neither do you need to learn this response because you inherited it when you were born. Other examples of reflex actions include coughing, swallowing, and sneezing. Notice that all normal human beings have

Spectacular feats of balance depend on perfect co-ordination between sensory messages and the brain's responses.

Instinctively, the long-tailed tit builds a complex nest.

the same reflex actions and other inherited responses.

The second type of inherited response is called a *taxis*. A taxis is movement towards or away from the direction of a stimulus. For example, fly-maggots move away from a light source. This is a normal reaction of all fly-maggots and is inherited.

The third type of inherited response is called a *kinesis*. A kinesis is a response to the presence or to the absence of a stimulus. Unlike a taxis, an animal does not respond to the direction of the stimulus. An example of a kinesis should make this point more clear. When a wireworm is placed in dry soil the worm responds by becoming very active and burrowing vigorously. It continues to burrow until it accidentally comes to wet soil, where its movements become slower. These responses are caused simply by the presence or absence of water.

The last type of inherited response is an *instinct*. The web-building behaviour of spiders is an example of an instinct. A spider knows how to build its web without having ever seen one, because web-building is a chain of reflex actions. The stimulus which starts the first reflex action is known as the *releaser stimulus*. All the other reflex actions follow automatically. Other examples of instincts include nest-building behaviour in birds and mating behaviour in sticklebacks.

Learned responses

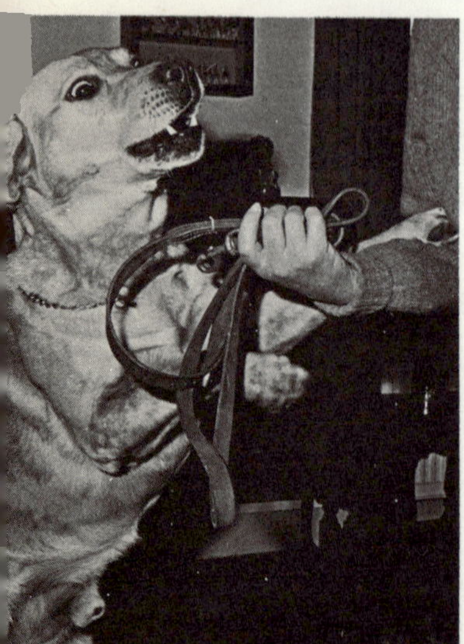

If you have a dog it probably becomes very excited when you get out its lead. When it was a puppy it probably did not show this response. During its life it has learned to *associate* (link) the sight of its lead with the pleasure of the walk which follows. Its behaviour, in this example, is known as a *conditioned response* which is a response aroused by some stimulus other than that which naturally produces it.

The response of a reflex action can also be conditioned. For example, a dog will *salivate* (its mouth will water) when food is placed in its mouth. This is a reflex action and it is natural and unlearned. Normally, the stimulus of a ringing bell does not make the dog salivate, but it can be made to do so. If a bell is rung when the dog is given food, and if this procedure is repeated many times, the dog will eventually begin to salivate at the sound of the bell alone. This conditioned response is an example of a *conditioned reflex*.

All animals that can learn anything at all can learn

conditioned responses. But not all learning is so simple. When a man learns a skill such as driving a car he learns a complicated series of responses. Skill-learning requires much training and practice.

Some animals learn response by *trial and error*. For example, an animal may find its way out of a cage by trial and error. When the animal is placed in the cage it may make several responses such as scratching the floor, banging against the bars, and touching the latch on the door. Only the last of these responses will solve the problem—the first two are errors because they achieve nothing. After the animal has been placed in the cage and has escaped several times it will make only the correct response. In other words it learns the correct response.

How responses are made

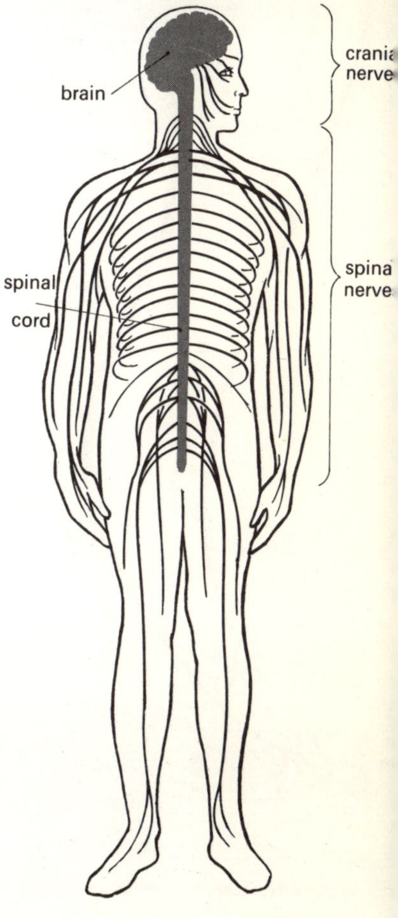

You have read how a stimulus can cause a receptor to produce nerve impulses (information messages) which travel along nerves to an animal's brain. (In fact, nerve impulses are sometimes sent to the spinal cord, but more about this later.) The brain and spinal cord make up the *central nervous system* (CNS for short). How does an animal respond to the information which reaches its CNS?

Think of the responses which occur when your CNS receives the information that you have put food into your mouth. You move your lower jaw to chew the food and your salivary glands produce saliva. Muscles are used to move the jaw, and glands are used to produce the saliva. Parts of your body, such as these muscles and glands, are known as *effectors* because they *effect* (carry out) responses. The body's effectors are joined to the CNS by nerves. The CNS sends nerve impulses along the nerves to the effectors. You can think of these impulses as *instruction messages* because they make the effectors start to work.

You should now see that nerves link both receptors and effectors to the CNS. First, you have 12 pairs of *cranial nerves* which link the brain directly to the receptors and effectors of the head. Second, you have 31 pairs of *spinal nerves* which link the spinal cord to the receptors and effectors of the rest of the body. Notice that messages sent between the brain and the body below the neck must travel along the spinal cord and spinal nerves. Messages sent between the brain and other areas of the head travel along the cranial nerves.

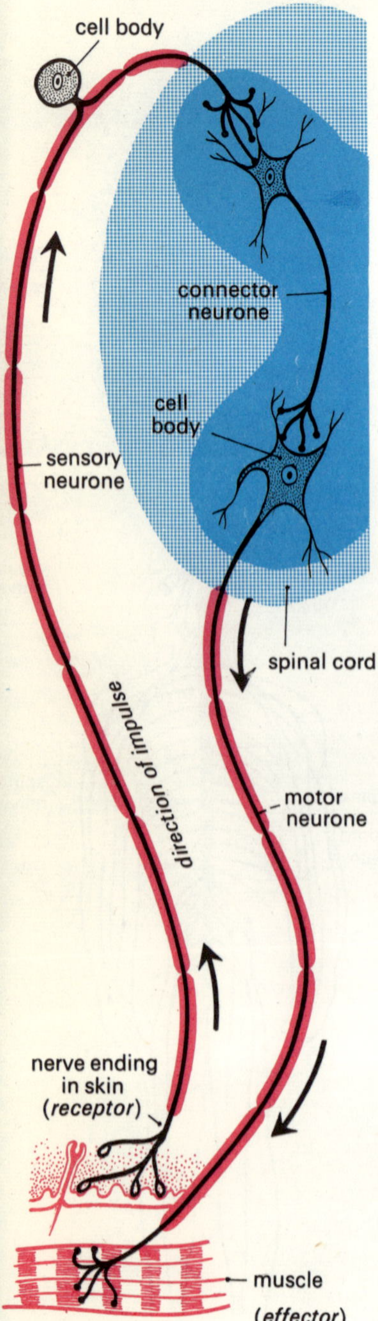

A reflex arc.

The nervous system—the brain, spinal cord, and nerves—is made of *neurones* (nerve cells). There are two types: *sensory neurones* and *motor neurones*. Both types have a cell body and a long fibre. A nerve is actually a bundle of nerve cell fibres, and some nerves contain both types of nerve fibre; others contain one type only. Sensory neurones carry information messages from the receptors to the CNS. Motor neurones carry instruction messages from the CNS to the effectors. Since each fibre carries its own message, there may be several different messages travelling through a nerve at any time. Let us now look in detail at how a particular response takes place. Study the diagram before you read on.

The illustration, shows, in a simplified way, what happens when your hand touches something very hot. A pain receptor in your skin produces impulses which travel along a sensory nerve fibre to your spinal cord. Next, the impulses pass to a motor nerve fibre which carries them to the biceps muscle in your arm. The muscle is an effector and the impulses cause it to contract. The contraction of the biceps muscle bends the arm and your hand is removed from the hot object. All this happens very quickly and automatically and requires no action by the brain. It is a reflex action, and the combination of receptor, sensory neurone, connector neurone, motor neurone and effector make up what is known as a *reflex arc*. This is, in fact, a *spinal reflex* because the spinal cord plays a part but the brain does not. Some other reflexes, such as blinking, are called *cranial reflexes* because messages are sent from the receptor to the brain and then to the effector. In cranial reflexes, as in all other reflexes, the brain does not make a decision.

What happens, however, when you *decide* to withdraw your hand from something? In this case, the brain itself produces impulses which travel along fibres passing down the spinal cord. These fibres connect with the motor neurone fibres in the nerve to your arm, so the impulses pass to the muscle which then contracts.

Nerve impulses

How do nerve impulses travel along nerve fibres? Refer to the diagram (right) as you read the explanation.

1 Look at the fibre coming from a receptor. The fibre is like a battery in the sense that it has a positive

(+ve) and a negative (−ve) pole. The outside of the fibre is +ve and the inside is −ve.

2 When a strong enough stimulus comes into contact with the receptor, the *polarity* at that end of the fibre reverses. In other words, the *inside* becomes +ve and the *outside*, −ve. This electrical change is caused by chemical changes in the fibre. The combination of chemical and electrical changes makes up what we call the nerve *impulse*.

3 These changes at the end of the fibre in region A start off the same changes in region B and so the impulse moves to B. At the same time, region A of the fibre returns to its original state. Thus, the impulse which was at the end of the fibre has travelled to B.

4 In the same way, the impulse at B travels to C and region B returns to normal. A new impulse now begins at the end of the fibre (region A) and follows the first one along the fibre.

By this method a stimulus causes a series of impulses to be sent along the fibre. The number of impulses formed in a second (the frequency of the impulses) depends on the strength of the stimulus. For example, a bright light causes more impulses per second than a dim light.

It may help you to think of a receptor as a coding machine. It is able to translate a stimulus into a code

When a receptor cell (left) receives a stimulus, the electrical charge between the inside and outside of the fibre changes and an impulse starts to move along the fibre (A), Before the next impulse is formed, the fibre returns to the resting state, and the electrical charge returns to normal.

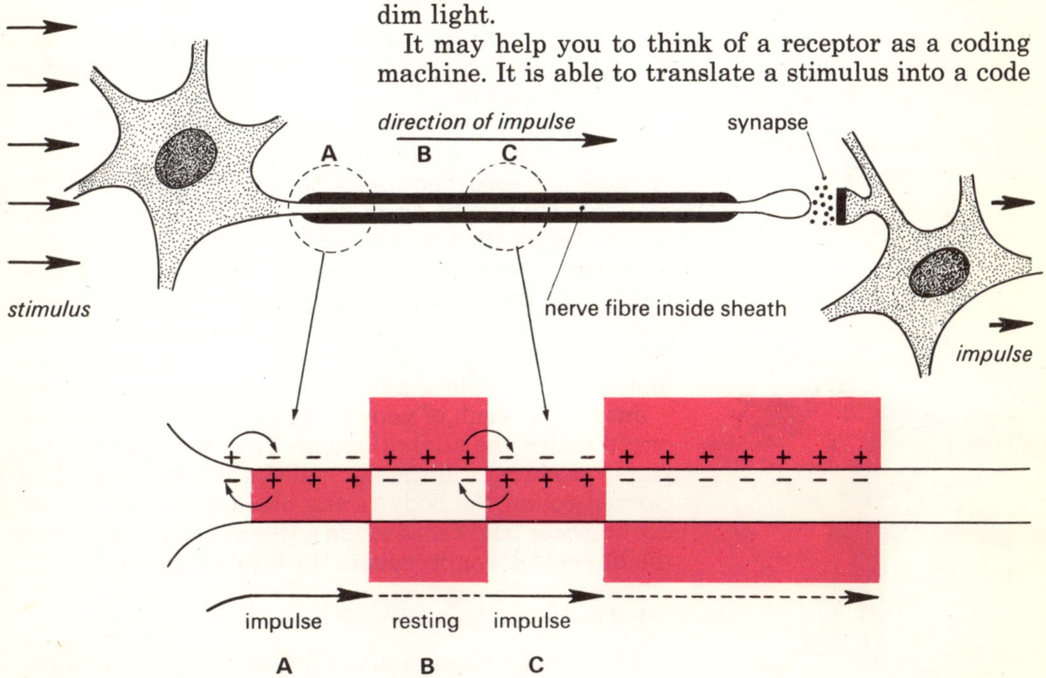

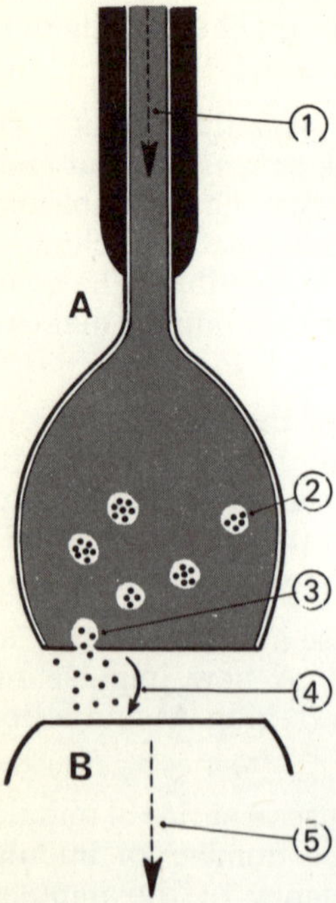

The impulse (1) causes the small sacks (2) at the end of fibre (A) to rupture (3), releasing a chemical into the gap (4). This causes the impulse to start up in fibre B (5).

composed of impulses. When the impulses reach the brain, they are translated into sensations. Although all impulses are exactly alike, no matter where they come from, the brain translates impulses from the eyes into pictures, impulses from the ears into sounds, and so on.

We have been talking so far about the impulses which make up information messages. However, the impulses which make up the instruction messages travel along fibres in a similar way. These impulses form coded messages which tell the effectors how hard to work.

An impulse may have to travel along more than one neurone in order to reach its destination. The diagram shows two neurones (A and B) which are linked. Notice that there is a gap, called a *synapse*, between the neurones. An impulse travelling along fibre A must cross the synapse before it can travel along fibre B. When the impulse reaches the end of fibre A a chemical is released and this spreads quickly across the small gap until it reaches neurone B. Here, the chemical causes a new impulse to be formed. The new impulse can now travel along neurone B.

Chemical messengers

You have seen how nerve impulses can cause an effector to make a response. But responses can be made by another method.

In the body there are many glands, and some of them —the *endocrine glands* or *ductless glands*—produce chemicals called *hormones*. Hormones pass from the glands which make them and into the blood and the blood carries them to all parts of the body. When the hormones reach certain parts of the body they cause responses there. In other words, hormones carry instruction messages coded in their chemistry and so they are sometimes called chemical messengers. Let us look at an example of a hormone and see how it works.

Just above each of your kidneys is a cream-coloured organ called the *adrenal gland*. The inner part of this gland produces a hormone called *adrenalin*, which is carried round the body in the bloodstream. When we are angry or frightened much adrenalin is released into the blood. Adrenalin causes the following responses:

1 The liver releases stored glycogen (a form of sugar).

2 Blood vessels expand in the heart, limbs, and brain.

3 The spleen releases its stores of blood.

4 The blood-pressure rises and the heart-rate increases.

5 The lung *bronchioles* (fine air-tubes) relax.

6 The pupils of the eyes dilate and the hair may stand on end.

Try to work out how each of these responses may help you to deal with a difficult or frightening situation.

Other hormones affect our behaviour in other ways. For example, the thyroid gland produces a hormone called *thyroxin*. A person becomes *lethargic* (drowsy) if his thyroid produces too little thyroxin. On the other hand, too much thyroxin causes overactivity and nervous tension. However, remember that there are many other causes of lethargy and overactivity.

In addition to the adrenals and thyroid, the *gonads* (testes and ovaries) and the *pituitary* gland affect our behaviour. The pituitary gland is especially important because it produces about ten hormones. It is sometimes called the master gland because its hormones control the activities of many other endocrine glands.

Physical stress, fight or fright are situations in which adrenalin is released into the blood to enable the body to take appropriate action, as shown above, below and opposite.

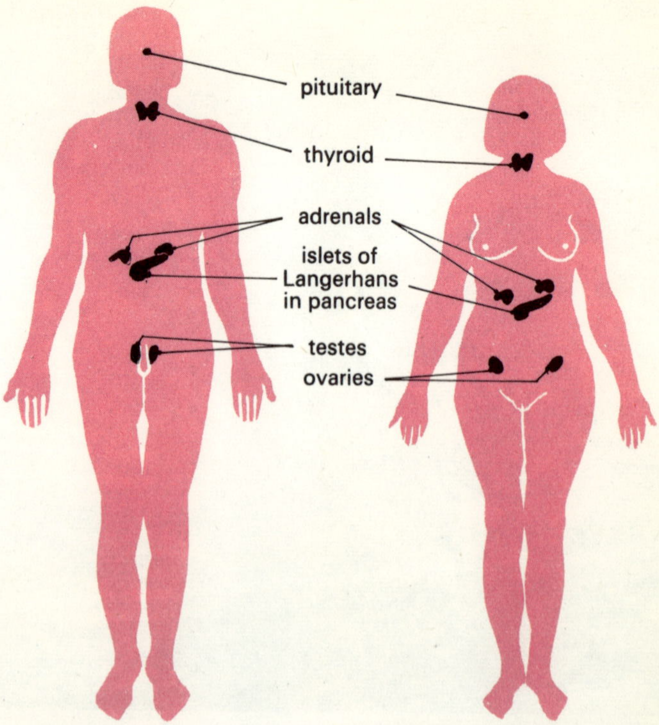

The endocrine glands.

Hormones take longer to produce responses than nerve impulses do. This is because impulses can travel along nerves at speeds over 100 m/s. Blood, which carries hormones, travels much more slowly.

Co-ordination of responses

Vesalius drew the muscles of a man in the sixteenth century.

When you walk you use fifty-four muscles to move one leg forward. To walk properly the muscles must act in the right order and each muscle must contract or relax by exactly the right amount. In other words, the muscles must be controlled so that they work together efficiently. This control is called *co-ordination*. The body is co-ordinated partly by hormones and partly by the spinal cord and brain. You have already read how adrenalin co-ordinates some parts of the body. It makes them work together so that they all help us to deal with an alarming situation.

The spinal cord helps to co-ordinate some responses. For example, a headless chicken is sometimes able to run just after it has been 'killed'. Its leg muscles are still co-ordinated even though they are not connected to the brain because the muscle movements form a chain of spinal reflex actions. You may remember that the brain is not essential for these reflexes.

However, the story is not quite as simple as this. Look at the diagram of the human brain, and you will see a part called the *cerebellum*. A man whose cerebellum is seriously injured is not able to walk normally. He moves in a jerky manner, which shows that his actions are not completely co-ordinated. Thus, the cerebellum in vertebrates is responsible for the fine control of body movements. How does it do this job?

It will help you to understand how the cerebellum works if you first think about the following situation. Imagine a man who is driving his car through a town and trying to keep exactly to the speed limit. To do so he will have to co-ordinate his use of the accelerator and brakes. This co-ordination is possible only if he can look at his speedometer and see the effects of his acceleration and braking. In other words, the speedometer informs him that he is co-ordinating the pedals correctly.

In a similar way, the cerebellum needs to *know* that it is co-ordinating the muscles correctly. This is possible in the following way:

1 The receptors in the muscles, tendons, joints, and ear-labyrinths send information messages to the cerebellum. These messages tell the cerebellum the position of the body.

This diagram of the human brain is 'exploded' to show what lies between and beneath the two cerebral hemispheres. The cerebral cortex can be seen where the rear lobe of one hemisphere has been cut.

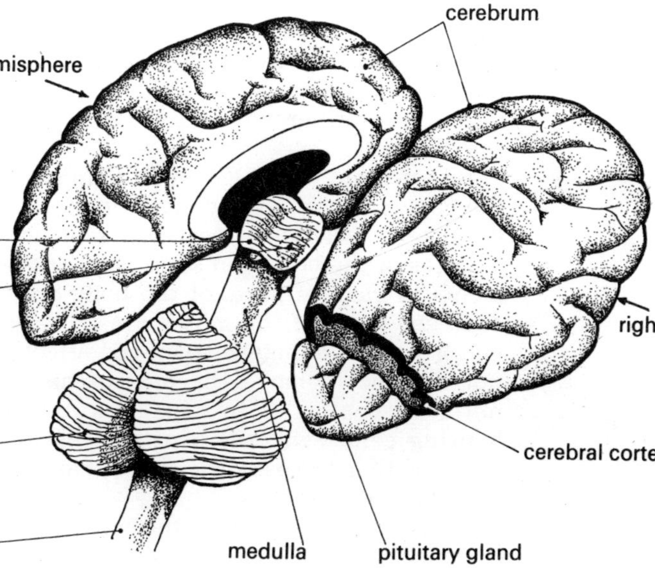

2 The cerebellum uses this information to co-ordinate the messages to the muscles to produce movement.

3 As the parts of the body move they send more information messages back to the cerebellum. These messages tell the cerebellum the new position of the body, and the cerebellum knows that it is co-ordinating the muscles correctly. Without this information the cerebellum would not know what it was doing and so it would make mistakes.

Responses and the cerebrum

The *cerebrum* is the largest part of the brain and it is divided into left and right halves called *cerebral hemispheres*. The cerebrum receives information messages from the receptors and sends instruction messages to the effectors. (It is an odd fact that the *left* hemisphere exchanges messages with the *right* side of the body and the *right* hemisphere with the *left* side of the body. This is why serious damage to the left hemisphere will cause paralysis on the right side of the body.)

The outer layer of the cerebrum is known as the *cerebral cortex* and it consists of the cell bodies of brain neurones. The cortex is grey in colour and is sometimes called *grey matter*, whereas the interior of the cerebrum consists mainly of neurone fibres and is *white*. Notice that the surface of the cortex is very wrinkled and that two of the large wrinkles (*fissures*) separate the cortex into lobes. By using several methods, scientists have been able to discover something about the functions of different areas of the cortex, as shown in the diagram.

Just in front of the *fissure of Rolando* is a portion of the cortex known as the *motor area*. It has this name because it is able to form and send impulses to the muscles of the body and so produce movement. Movements in various parts of the body are controlled by different parts of the motor area. If a portion of the motor area is damaged, then the affected person will not be able to move the part of his body which is normally controlled by that area. Sometimes, neighbouring cells will take over the job of the damaged area and then the person will once more be able to make voluntary movements.

There are other areas of the cortex which are known as *sensory areas* because they receive impulses from the body's sense receptors. To the rear of the fissure of Rolando is the 'body feeling' area, which receives messages from the skin and gives us our experiences of temperature and touch. This area also receives impulses from the moving parts of the body and gives us experiences of movement.

Impulses from different parts of the body arrive at different parts of this sensory area and enables us to experience the sensation in the particular part of the body which has been affected. In addition to this body feeling area there are sensory areas which receive impulses from the eyes and ears and produce the experiences of sight and hearing. The visual area is in the *occipital lobe* and the *auditory* (hearing) area is in the *temporal lobe*. There are also sensory areas concerned with taste and smell but no-one is very clear about how these work.

We have seen that the cortex has motor areas which produce impulses sent to the body and sensory areas which receive impulses from the receptors. In addition to these areas, there are other parts of the cortex which are known as association areas because they connect and integrate information which reaches and is stored in the brain. Some association areas are involved in the abilities to understand the meaning of both written and spoken words, and the abilities to speak and write meaningfully. Other association areas in the frontal lobes of the cortex are believed to be used in abstract thinking.

Left: some of the functions of the cerebral cortex. Areas of the body which are very well supplied with sensory receptors, such as the hands and face, are represented by large areas of the cerebral cortex. It is thus possible to construct a 'homunculus', or human figure which shows by its distortion which parts of the body have the most nerve endings.

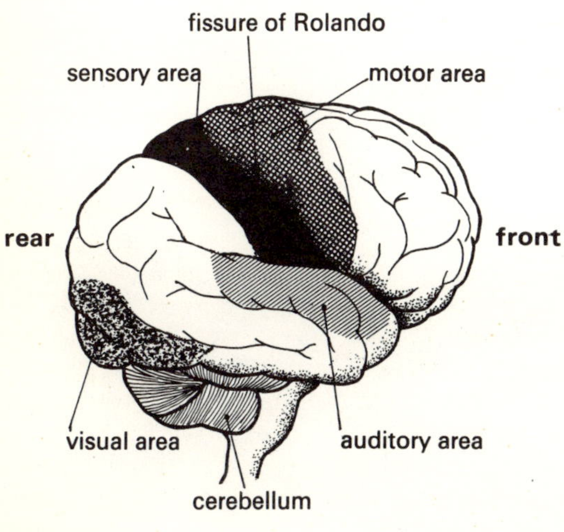

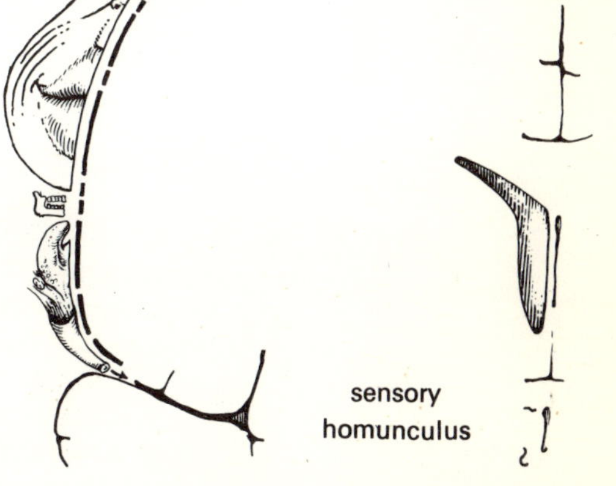

3 What makes us behave as we do

You have read how a stimulus from your surroundings can produce a response; in other words, a stimulus can cause a change in your *behaviour*. However, stimuli coming from your own mind and body also affect the way you behave. Let us look at some of these.

The needs of the body

If you are to survive you must satisfy your bodily needs such as the need for water and food. When your body is lacking one of its needs you begin to feel uncomfortable and this discomfort acts as a warning. For example, when your body lacks water, your throat and mouth become dry. This warning will make you behave in a way which will remove the feeling of discomfort and also satisfy the need. In this example you would look for liquid and drink it.

Other bodily needs include the requirement to take in air, to rest, to be active, to sleep, to urinate and defaecate, to escape danger and to avoid pain. Think how much of your behaviour is concerned with satisfying these needs and also try to work out the warning signs of each need.

We are all born with bodily needs and they stay with us throughout our lives. As we grow older, we learn from other people the acceptable and unacceptable ways to satisfy our needs. For example, we learn to eat our meals at definite times in definite places. You will be able to think of the acceptable ways of satisfying other bodily needs.

In extreme conditions the body makes urgent demands, and the responses may seem strange. It helps the athlete to lose heat when he runs if he sprays his face with water from a sponge.

When a person is unable to satisfy a need it may have

an extreme effect upon his behaviour. For instance, if a man cannot satisfy his thirst he will think more and more about water as he gets thirstier. Eventually he will be unable to think about anything at all except water. Our descriptions of such a person as 'crazy with thirst' show how much his behaviour has become affected.

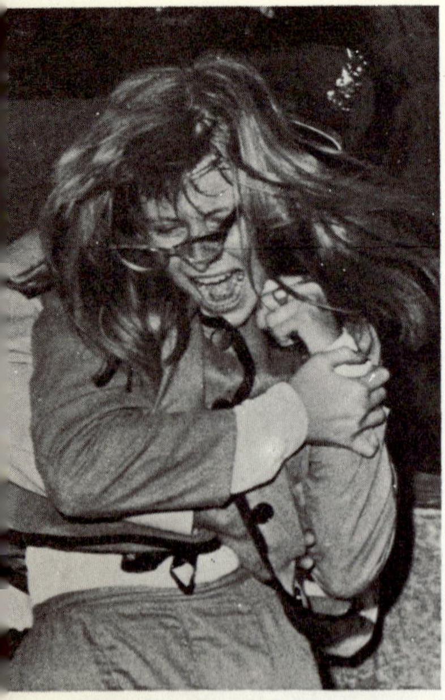

You can tell that the girl in this picture is feeling strong emotion, but what does her face reveal? Is it distress as she is manhandled by a police officer or ecstacy caused by the pop concert she is attending?

Feelings and emotions

We all know that emotions such as fear, anger, joy, grief, guilt, love and hate play a very important part in our lives. We can often recognize the emotion a person is feeling by watching his behaviour. Smiling, crying, and banging a fist on a table are all bits of behaviour which indicate different emotions. Emotions are accompanied by changes inside the body. For example, our hearts beat faster and we breathe faster when we are afraid. We are not sure how the brain works in creating our emotions but it seems clear that the cortex and the *hypothalamus* (see page 33) play important parts.

Human beings often suffer emotions such as fear when there is no real cause. People imagine that something terrible is going to happen and then feel afraid. Emotions which are caused by real or imagined reasons can sometimes make people ill. Such illnesses can cause skin disorders, migraine and ulcers, although you should remember that there are other causes for these illnesses apart from emotional ones. It should be apparent by now that both the mind and the body are affected by emotions.

As you know, some emotions are pleasant and some are unpleasant. Also, we can experience an emotion in various states of intensity: someone may feel slightly annoyed or very angry. When a person is feeling an intense emotion he will often pay a lot of attention to the cause of the emotion and is more likely to become active as a result. An angry person thumps the table and stares wild-eyed at the cause of his rage.

Emotions can either help or hinder us. For example, in an emergency a certain amount of fear may help to give us extra strength to deal with the crisis. The liver releases more sugar into the blood and gives us a greater source of energy. On the other hand, if we are too frightened we may become 'rooted to the spot' and incapable of any action.

Intelligence

Intelligence plays a large part in determining how we behave. Intelligent people will be able to do things which cannot be done by less intelligent people. Most of us are fairly sure that we can recognize intelligent behaviour but, in fact, scientists still argue a great deal about what is meant by intelligence. Some believe that intelligence should be measured by how well we cope with (or learn to cope with) situations which are new to us. Others think that it is better to measure intelligence by how well we do tasks in which we need to understand the relationships between things—in other words, how well we can think in abstract terms.

After a person has been given an *intelligence test*, the results can be used to calculate that person's *IQ* (*intelligence quotient*). A person with an IQ of about 100 has average intelligence. A small number of people have IQs below 70 and are often mentally retarded. A few people have IQs of more than 130 and are very intelligent.

Very little is known about how brain cells function in intelligent behaviour, but we do know that the cells of the cerebral cortex play a most important part. If you compare a man's brain and a dog's brain and you can see that the proportion of the cerebrum to the rest of the CNS is greater in the man's brain. It is this relatively large and highly developed cerebrum which makes a man more intelligent than a dog. Brain growth

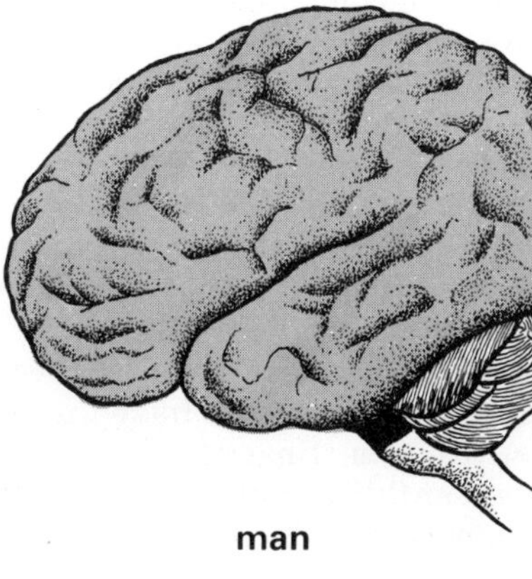

There is a definite link between the surface area of the cerebrum and intelligence. Thus it is not size that counts, but wrinkles.

Next to man, the dolphin has the most highly developed brain. Thus it is capable not only of learning tricks but also of performing complicated tasks which require a high degree of intelligence.

may be arrested. People with brains of less than 800 g are usually feebleminded, although there is no clear correlation between brain size and intelligence.

Learning and memory

Generally, the way we behave is influenced by our characteristics at birth and our experiences during life. Since no two people are exactly alike at birth and no two people have exactly the same experiences, it is not very surprising that no two people behave in exactly the same ways.

The things we learn can be simple or complicated. For example, a young child may accidentally put his hand in boiling water and learn that boiling water is dangerous. On the other hand, we may deliberately learn something complicated such as mathematics. These two kinds of learning have one fact in common: they both cause changes in behaviour. Once someone has put his hand in boiling water he is unlikely to make this response again. Similarly, after he has learnt mathematics he will respond to mathematical problems in new ways.

All of us accidentally learn many things during our lives, but when we deliberately try to learn something, such as mathematics, some of us are more successful than others. Why is this so? Intelligence plays a part. The more intelligent we are, the more easily we are able to learn. People who are emotionally disturbed, however, may find concentration and learning difficult. People seem to do better if they have a reason or a *motive* for learning. The motive may be simply an interest in the subject or it may be the desire to pass an examination or receive praise.

What we learn must be remembered if it is to influence future behaviour. Learning causes changes in the cerebral cortex and these changes make up our *memory*. No one knows exactly what these changes are but one theory is that learning causes chemical changes in a group of brain cells which then become connected together to form a memory chain or circuit. This circuit of cells becomes active when we remember what was learned.

The difference between what we have learned and what we have remembered is what we have forgotten! We all tend to forget things as time goes by. However, because we can remember things which happened years ago and yet forget something which happened yester-

day, time cannot be the only cause of memory lapses. Our memory of a fact may be interfered with by what happened before and after we learned that fact. For example, if you went to the zoo with the idea of learning the names of a large number of animals you would probably have difficulty in remembering them afterwards because the names you tried to remember at the end of your visit will be interfered with by the names you tried to remember earlier on. Similarly, the later names will interfere with the earlier ones.

Cookery classes are popular with primary school children. Not only will they take part in the making, but they are also motivated by the prospect of eating.

Acquired motives

Each of us has needs such as the need for water or the need to succeed. These needs cause us to behave in ways which will lead to their satisfaction. A motive or a drive are other names for such a need. You have already read how we are born with certain body needs such as those for food and water. However there are other motives which we acquire during our lives.

Some of these acquired motives are called common social motives because they are found in nearly all human beings. One such motive is *gregariousness* (the desire to keep in contact with other people). As children, we all go through a stage when we are helpless and dependent on other people. For this reason, we learn to accept the company of others. Through contact with others we learn the need to imitate others and to seek help from other people when we cannot cope with a situation alone.

During our lives we acquire personal motives, also. Our personal motives can vary a great deal. For example, one man may feel the need to be admired by a few acquaintances. Another may aim to become a pop star or film actor to be admired by a much larger number of people. Both are variations of a common need to be recognized. People can have different motives for aiming for the same result. Thus, people may wish to become politicians either because they want to gain power over others or be admired by others, or because they want to be in a situation where they can help others.

Personality

When you meet someone you probably soon form an opinion about his personality. You may describe the person as having a 'lively', 'dull' or 'interesting' personality. In other words, you form an overall impression of the person. However, if we want to describe personality more accurately we must look in detail at every aspect of a person's make-up. You have read in this chapter about emotions, intelligence and motives, and these are just a few of the many aspects of personality.

Sigmund Freud (1856–1939), a famous Austrian psychiatrist, came to believe that personality consists of three parts which he called the *id*, the *ego* and the *superego*. The id is that part of the personality present at birth. It consists of desires to satisfy biological (body) needs and to avoid pain. As a child grows older he builds up a picture of himself and his surroundings and learns acceptable ways of meeting the demands of his id. This part of personality is called the ego. The child will also gradually learn what is right and wrong. That is, he develops a conscience or superego.

In an adult, the ego acts rather like a referee between the demands of the id and the superego. For example, if you are hungry and see food in a shop, the id would demand that you grab the food and eat it, but the superego would tell you that this is wrong and you must not do so. The ego then has to satisfy the id and the superego by finding some socially acceptable way of obtaining food.

Other psychologists have tried to divide people into two personality types called *extraverts* and *introverts*. Extraverts are people whose behaviour is directed, in the main, towards other people. They enjoy other peoples' company. Introverts are people whose activities tend to be directed towards themselves. They prefer to be alone. It is now accepted that most people are *ambiverts*, which means that they are extraverts in some situations and introverts in others.

Attempts have been made to find a list of characteristics (or *traits* as they are known) which would be used to describe a personality. Examples of such traits are dominance, sociability, intelligence, self-control, and so on. A person can be rated for each of these traits and we can then decide how dominant, how sociable, how intelligent, and so on, that person is. Once this has been done we can describe his personality according to his ratings in all the traits. This is a useful way of roughly describing a personality but we should remember that each of us has traits which are found in no one else. We return to the fact that each human being is unique and that the reasons for human behaviour are very complicated and still not well understood.

What can you see in this shape? The Rorschach inkblot is well known in personality tests. Try making inkblots of your own between pieces of paper and compare interpretations.

4 Couples

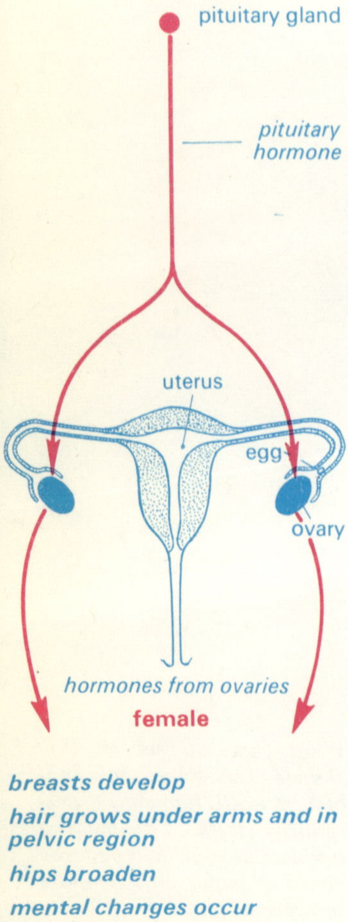

So far you have been learning how animals respond to information from their senses, and how these responses help animals to survive. You have also learned a little about the ways in which human behaviour can be influenced by bodily needs, emotions, personality, and so on. What you have read so far concerns the behaviour of animals as individuals. We have said little yet about the behaviour of one animal towards another, or about the behaviour of animals in families, groups and societies. Yet, few animals live their lives entirely alone. They must, at some stage, come together in pairs at least to reproduce so that the species will survive when they, the individuals, die.

Sexual maturity

Mammals cannot reproduce as soon as they are born. They undergo a period of growth first. Young animals need all their food to build their bodies as they grow. If young females were to reproduce early, their bodies would be too frail to carry the young and their bones would be too small to allow the young to be born.

At some time after birth, however, the sexual organs become active in response to hormones from the pituitary gland. The gonads enlarge and eggs and sperm are produced. There is some evidence that, in girls, this process begins when they reach a certain body weight. The gonads themselves begin to produce hormones, too. Male gonads produce *androgens* and female gonads produce *oestrogens*. These are the hormones which bring about body changes such as the growth of facial hair in boys and the broadening of the hips in girls.

Maturation and behaviour

Sex hormones not only bring about body changes, they affect behaviour too. Mature animals possess a *sexual drive*. You have seen that drives, or motives, are responses to an animal's basic needs. The sexual drive differs from others because it is important to the survival of the species but not the survival of the individual. In other ways it is similar to other drives, however, because it produces a typical pattern of behaviour—a period of searching, an achievement, and then a period of rest.

Sexual rhythms

Very few animals remain sexually active throughout the year. The level of sex hormones in their blood varies during the year, producing short periods of sexual activity separated by long periods when the sexual drive is absent. These alternating periods of activity and inactivity do not occur at random. They tend to follow a regular pattern. Such regular patterns of animal behaviour are called *rhythms*.

The sexual rhythm may be under the control of some feature in the animal's environment. For example, the gradual increases in the hours of daylight during spring cause the gonads of birds to enlarge. This enlargement, in turn, increases the amount of sex

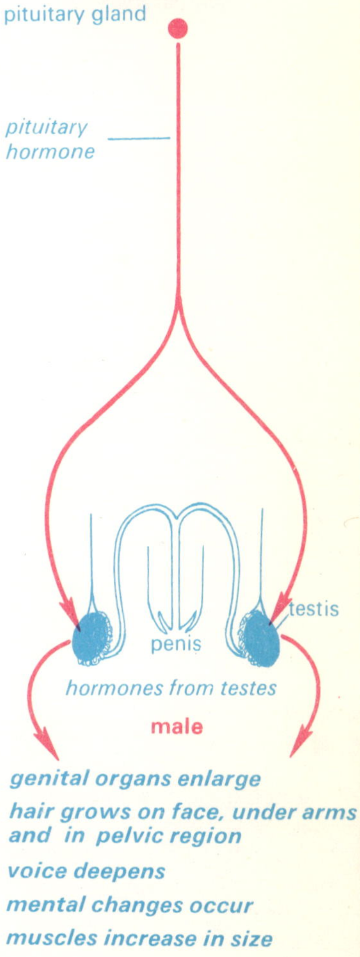

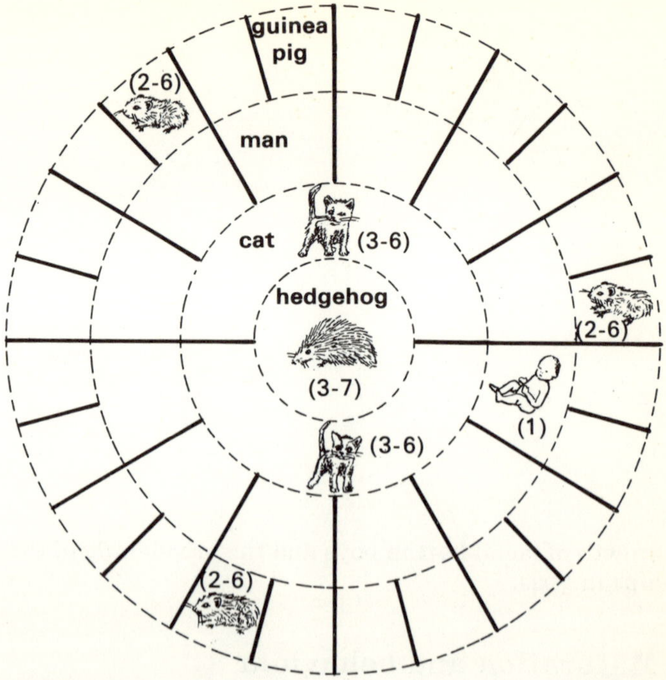

In the diagram a complete circle represents one year. The guinea pig produces eggs twenty-four times in a year, though on average it will only have three litters of between two and six young. Ovulation in humans occurs twelve times a year, in the cat twice, and in a hedgehog only once. Below: sex hormones in the blood cause plumage changes in the male eider duck (bottom). The female (above) responds to his plumage.

hormone circulating in a bird's blood and awakens the sex drive. Although this process appears to be controlled mainly by increasing day-length, it has been shown that in pigeons and other birds a female's ovaries will develop more quickly if she is kept with a male of the same species.

Communication

Animals which are ready for reproduction must be able to 'tell' their partners about it—and their partners must be able to respond, as well. An exchange of information before sexual reproduction is one of the commonest forms of communication between animals. In many species this exchange has developed into a series of inherited actions and responses called *courtship behaviour*.

Finding a mate

Some animals live in herds and, for them, finding a mate is easy. Other animals spend their lives away from members of the same species. How do these animals find their mates? Instead of wasting time in searching, many animals produce signals which enable their mates to find them.

Female flour moths parade with scent glands extended to attract the male.

Gipsy moths, and other moths, use chemicals called *pheromones* to help their mates to find them. Animals release pheromones into their surroundings; these are sometimes called *social hormones* because they affect the behaviour of *other* animals of the same species whereas ordinary hormones affect the behaviour only of the animal which produces them.

A female gipsy moth releases tiny quantities of pheromone into the air. A male can detect the faintest trace of pheromone on his feathery antennae up to 3 km away. He responds by flying upwind, bringing him steadily nearer to the female.

Sensitivity to pheromone has been used to lure many male moths to their deaths. Gipsy moths are pests; they cause serious damage to fruit crops and, in an effort to reduce this damage, scientists have isolated chemicals similar to the pheromone produced by the female. Scientists release the chemical into the air and male gipsy moths fly upwind to the point of release, where they are captured and destroyed. Removing the males from the gipsy moth population is an effective method of control because it obviously prevents reproduction.

Other animals make use of chemical signals, too. Even in man the sense of smell is still important. We wash our bodies carefully to remove our own natural scent, only to replace it with commercial perfumes which we expect to make us more attractive to the opposite sex. This practice is not restricted to women, although men usually describe their scents as after-shave lotion, and so on.

Other animals make use of more familiar stimuli. Male crickets use sound to attract their mates. French crickets rub their wings across their backs to produce a characteristic noise. This is *stridulation*. Female crickets move towards the sound. Other animals, including birds and toads, use *assembly calls* to attract females.

Many animals locate their mates by sight. The bright coloured patterns on butterflies' wings serve this purpose. There is even one night-flying beetle which finds its mate by sight. This is the *firefly*. The female has no wings; it attracts the flying male with flashes of light, and each species has its own pattern of flashes.

The pheromone of the gipsy moth, the stridulation of the cricket, and the flashing light of the firefly are all stimuli which make members of the same species act in a particular way. Such specific signals are called

Below the stag's eye is a 'scent gland'. By rubbing his head on trees he leaves a scent on the bark as a signal to other deer. Peacocks and fireflies use signals which can be seen. Grasshoppers and crickets attract their mates by sound.

The West African clawed toad first extends the air sac under his chin, and then uses it as a resonator to make his sexual signals louder.

releasers. Releasers can carry a lot of information. Not only can they show where an animal is, they can also give details of its species and its sex.

Animals are bombarded constantly by a variety of stimuli, but the brain 'filters out' most of them and reacts to only a very limited number of important stimuli. These are called *sign-stimuli*.

How do animals ensure that they meet and breed with animals of the same species? Courtship behaviour must try to prevent interbreeding between animals of different species. Mating between animals of different species rarely gives young, and when it does the young are often *sterile* (unable to breed). For example, cross-breeding a horse and a donkey produces a sterile mule. The object of reproduction is to ensure survival of the species and the production of *hybrids* (cross-breeds between species) does nothing to help this. As a result, many animals have developed patterns of behaviour called *reproductive isolation* to avoid interbreeding.

In North America, two species of frogs have almost identical assembly calls. One species lives mainly in the east, and the other mainly in the west. However, in the middle of America, these two species share the same ponds and in this area their assembly calls are entirely different.

Overcoming aggression

Appeasement.

Outside the breeding season, many animals react aggressively if they meet other members of the same species. One of the jobs of courtship is to overcome this aggression.

In mammals and birds, females often avoid attacks from aggressive males by patterns of behaviour called *appeasements*. In birds, it is quite common for a female to act like a young bird begging for food. Female black-headed gulls do this. It seems to make the male less aggressive, probably by stimulating his parental behaviour.

Co-ordination

When male and female have been brought together they still have to co-ordinate their activities so that reproduction can take place. Their natural tendency to avoid each other must be overcome. Sometimes this problem is solved in the most direct way. Having collected a number of female frogs with his croaking assembly call, the male simply clasps the nearest female. (He may clasp a male just as readily, but he will be 'warned off' by a croak!) However, many animals have a complex pattern of behaviour which they must complete before they reach the sexual act. The pattern may consist simply of laying eggs and releasing sperm, or it may be an act of copulation involving physical contact between the two animals.

The queen butterfly offers a good example of a complex pattern of reproductive behaviour. If a male butterfly sees a female overhead he flies after her and overtakes her. He then extends two bunches of hairs called *hair pencils* from the end of his body and dusts these round the head of the female. The hair pencils are rich in a strong scent. In response to the scent, the female lands. The male then hovers overhead and continues to dust the female until she closes her wings. In response to this wing-closing, the male lands alongside the female and mates with her.

In the project section at the end of the book you will find details of an investigation which you can carry out in the laboratory into the courtship behaviour of fish.

Summary

You can see that courtship is a complicated process. A sexually mature animal may be brought into breeding condition by a change in the length of the hours of daylight. It must then find a mate somewhere in its surroundings and identify it as being of the correct sex and species. Any aggressive tendency of the potential mate must be overcome. Finally, the two animals must co-operate to make possible the fertilization of the eggs. In mammals, birds, and reptiles, physical contact between the sexes enables sperm to be transferred.

Courtship may be complicated further by competition, fighting, and territorial behaviour, all of which are discussed later in the book. However, courtship behaviour does improve the chances of successful reproduction. It also serves to prevent interbreeding because courtship behaviour is *species-specific*; in other words, it produces responses only from members of the same species.

mewing

choking

facing away

Three of the many courtship postures recognized by gulls.

Human courtship

How does human courtship differ from courtship in other animals? In all the examples we have considered so far, courtship has been an *innate* (unlearned, instinctive) pattern of stimuli and responses. Human courtship is not so simple. Some aspects are probably innate, but they are greatly altered by courtship patterns we learn from the group in which we live.

How much of our courtship behaviour is innate is not yet clear. We have still got a lot to learn and research is difficult because many of the signals we use seem to be so subtle and we may respond without knowing why. For example, in an experiment some men were shown two pictures of a woman's face which, at first sight, appeared to be identical. Yet when the men looked at one of the pictures, the pupils of their eyes enlarged more than when they looked at the other. None of the men knew that this was happening. When they were told about it they could not understand why, because both pictures looked the same. What they had not realized was that, in one of the pictures, the woman's pupils had been enlarged. But why did the large pupils in the photograph cause the men's pupils to enlarge? The investigators believe that when we see something which interests us, our pupils get bigger. Thus, the men in the experiment were more interested in one of the photographs than in the other. What is more, they were more interested in the photograph with the enlarged pupils—that is, the photograph of the woman who appeared to be showing more interest in them!

Eyes and lips are used in signalling and courtship. Make-up emphasises them. The girl alongside has no make-up.

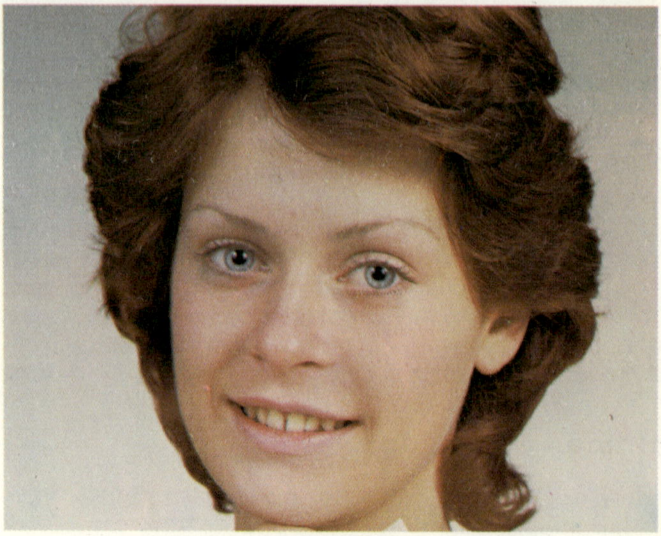

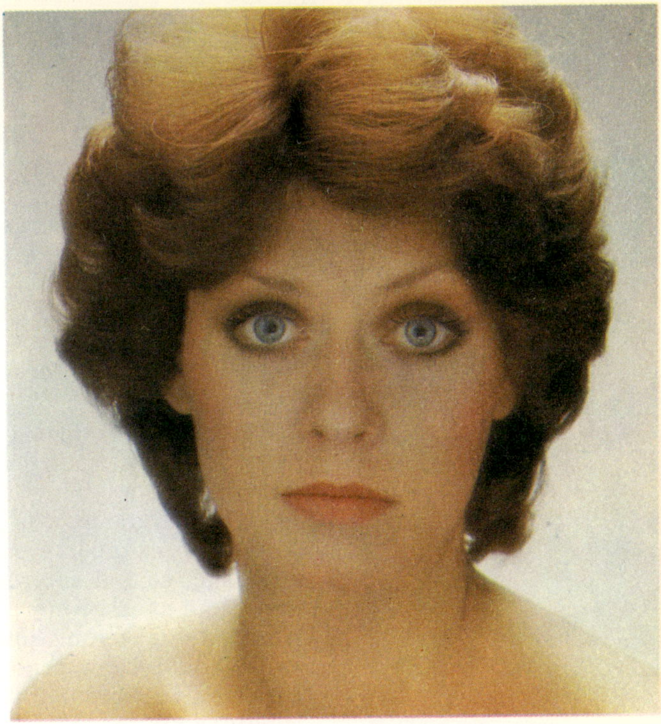

This is the same girl. Do you think she looks more attractive now?

Throughout the world, scientists have filmed people without their knowledge. These films have been analysed and it has been claimed that there are certain facial expressions which are common to people of many races. It may be that these expressions are inherited. The basic facial expressions of flirting—an early stage of courtship—are said to fall into this category. Flirtatious facial expressions of girls from many countries follow the same basic pattern. The girl will first smile in a welcoming way, then lower her eyes and turn her head away, as though to indicate that she is leaving. This pattern may be repeated several times.

After initial signs of interest, courtship usually continues verbally. The actual words used will vary with our background and experience, but they will usually be spoken in a soft and gentle way. Verbal courtship is followed by slight physical contact and, as courtship proceeds, this contact becomes closer and more prolonged.

In the courtship ritual, animals use a variety of aids. The male may become brightly coloured and even carry presents to his potential partner. What aids do we use? Look at clothing through the ages. Is it designed simply as protection or as a means of attracting members of the opposite sex? Why do women use cosmetics? What facial features do cosmetics emphasize? Can you think of any reasons for this emphasis?

Look back over the courtship behaviour of other animals. How many of these features do you think are fulfilled by human courtship? Are there any features of human courtship which make it different from that of other animals? Which aspects of our courtship patterns do you think are learned from the group in which we live? Can you find out anything about courtship in other periods of our history or in other parts of the world?

Human courtship may lead to a marriage lasting forty or fifty years. So human courtship should do more than enable a young couple to achieve successful sexual intercourse, which is not usually very difficult. Although sexual happiness is important to a happy marriage, sexual intercourse only occupies a small part of an average married couple's life. What you do with the rest of your time is equally important. Courtship should be a time for finding out about your partner's background, interests, and temperament. It should show areas of serious disagreement between you. It may reveal that your partner has an uncontrollable temper, or is mean and selfish, or is simply very boring.

The sex drive is very strong and unless you use care and self-restraint it can lead to sexual intercourse without any forethought or planning. This, in turn, may produce an unwanted pregnancy. There are three solutions to this problem—abortion, an illegitimate child, or a hasty marriage. Which of these solutions do you think is the most satisfactory? Is there any other alternative?

It is the tradition for young people to marry soon after puberty in eastern Nepal. The bride is 16, her bridegroom 14, and they wear rice grains on their foreheads at their wedding as a symbol of fertility.

Traditional Zulu weddings are accompanied by animal sacrifice and a tremendous amount of ritual. The bride's family give a dowry, which was once obligatory in European societies too.

Pair formation usually depends upon the attraction between a male and a female, but in some cases two men or two women may be sexually attracted to each other. These homosexual relationships are also found amongst animals.

The causes of homosexual behaviour are not clearly understood. They are probably partly inborn and partly acquired. The Freudian psychologists believe that we go through a homosexual stage during our sexual development. It is only a few years ago that male homosexuality ceased to be a legal offence in this country. But today's homosexual still faces a great deal of ridicule and abuse. This is unfortunate since homosexuals are as sincere and no more 'wicked' than other people.

You have seen in this chapter that the sex drive brings animals together and requires them to communicate. Much of this courtship behaviour is in-inherited although, in man, inherited patterns are modified by the influence of the group in which people live.

5 Parents and Young

The more care an animal gives to its young, the fewer eggs it needs to produce for its species to survive. A thrush only needs to lay five; a turtle must lay several hundred. The cod will produce enough eggs to fill a thousand pages like the one opposite, and only two may reach maturity.

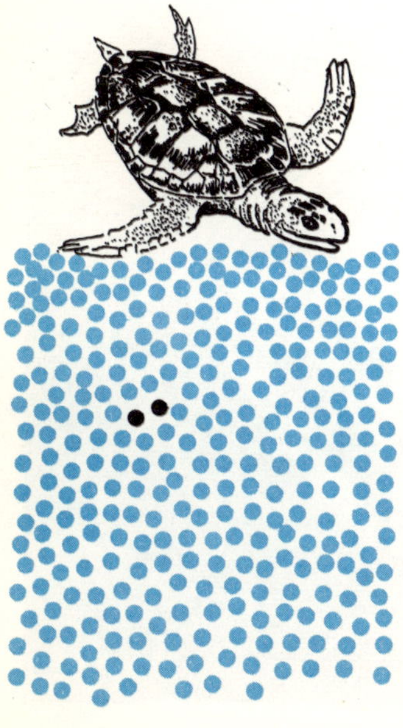

Courtship and reproduction create young animals to replace older individuals as they die. But this is not the whole story. Fertile eggs or young animals do not guarantee that the species will survive. To make certain of survival some of the young must mature to a point where they themselves can reproduce.

Animals have developed many methods of making sure that some of their young will reach the adult state. Some species, such as the cod, produce very large numbers of young, so that even if thousands die or are killed, a few will survive. Between January and April, in the Northern Hemisphere, cod gather into shoals and move into waters between 60 m and 100 m deep. Here eggs are released and fertilized. The eggs float to the surface where they hatch after 9 to 20 days. The eggs and *fry* (young fish) may spend up to 14 weeks as part of the *plankton* (the drifting community of plants and animals in the surface layers of the seas). Throughout this time, they are eaten in vast numbers by fish, jelly-fish, arrow worms, and comb jellies. To counteract this huge death-rate, cod produce astronomical numbers of eggs. Females in the shoal lay between 3 million and 7 million eggs *each*!

At the other end of the scale are animals which produce small numbers of young and take great care of them. A human mother may produce only one child, but she takes great care of it for fifteen to eighteen years.

Between these two extremes the animal kingdom shows great variety in both the number of young produced and the amount of care which they are given. Alligators care for their eggs only up to the point when they hatch. Sticklebacks remain with their newly hatched offspring for about a day. In birds and mammals the period of care is much longer. It is true to say in general, however, that the lower the number of young, the greater the care they receive. The need for greater care of the young means that at least one of the parents has to remain with the eggs after they are laid and hatched, or to stay with the young after they are born.

So survival of the species can be ensured either by producing very large numbers of young or by producing fewer young and taking greater care of them. But young animals are not cared for only because they are produced in small numbers. Some animals need care for a long time after they are born because they are totally unable to take care of themselves.

What do we mean by care? What advantages do young animals gain from the continued presence of their parents? Before you read on, make a list of as many advantages as you can think of. We will consider a few of them in detail.

Shelter

By shelter, we mean protection from extreme conditions. Eggs and young animals are sensitive to extreme heat and cold, dryness, and shortage of oxygen.

The male moth passes sperm directly into the body of the female, where fertilization takes place.

Animals which live in the water obviously never dry out, and temperature variations are small. *Terrestrial* (land-living) animals face many problems, however. *Desiccation* (drying out) is a constant threat, and seasonal and daily temperature changes can be large.

Sperm and eggs are especially prone to desiccation and need protection if they are to survive. Thus, frogs, toads and newts, which spend a large part of the year on land, return to the water to breed. The female lays her eggs in the water and the male releases a cloud of sperm which swim to the eggs and fertilize them. In reptiles, birds, and mammals, fertilization is *internal*: sperm are released by the male directly into the reproductive system of the female. In this way, they avoid contact with dry air and remain at a steady temperature.

After fertilization, eggs develop into embryos which require shelter, too. Shelter can be provided by laying the eggs in suitable surroundings. Fishes' embryos are protected from desiccation by the water in which they

The toad clambers on to the back of the female as a signal for her to release her eggs in black strings. The male immediately releases the sperm over them, for fertilization to take place outside the female's body.

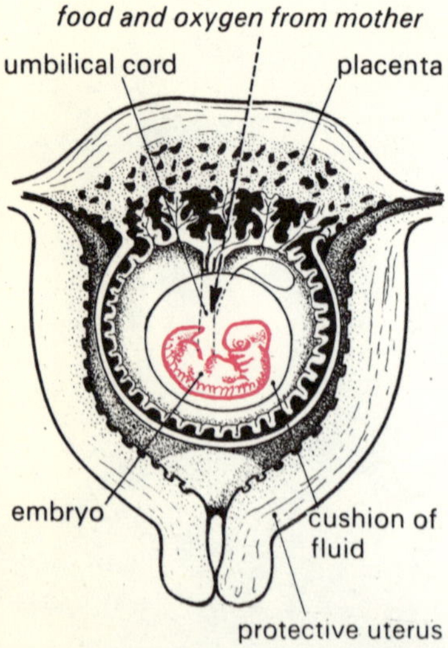

The human embryo inside the womb.

Balancing the single egg on his feet, and keeping it warm with his stomach, the male Emperor penguin may have to fast for three months or more as he broods the egg, while his mate is at sea feeding. She returns before the egg hatches, and feeds the chick on regurgitated food.

hatch. House-flies lay their eggs in moist surroundings —often in decaying meat—so that the larvae hatch out into comfortable surroundings. Embryos may be sheltered by surrounding them with a suitable cover. Birds' eggs have chalky shells around them. Reptiles' eggs have leathery coverings. The developing caterpillar of a butterfly is also sheltered by its egg. In birds and reptiles, the embryo is protected from desiccation by the watery *albumen* (egg-white) which surrounds it. Mammals spend the first part of their lives inside a womb. The womb is a wet, padded envelope kept at a constant temperature; it gives the embryo the best possible shelter.

In birds and reptiles, not only does the egg shelter the embryo, the eggs themselves are sheltered by the parents in a variety of ways. Many reptiles bury their eggs. Green turtles spend most of their lives at sea but lay their eggs on land. A female green turtle will swim hundreds of kilometres to reach a certain beach. Once there, it pulls itself ashore to dig a hole in the warm sand in which it lays its eggs. The female covers eggs with sand and returns to the sea. Some snakes bury their eggs in piles of decaying plant material. The eggs are protected from desiccation, and kept warm by the heat released during the process of decay.

Birds have a different method of sheltering their eggs. One or both of the parents incubate the eggs by sitting on them to keep them warm. Many birds shelter their eggs more efficiently by building nests. This is particularly true of perching birds which live in trees and hedgerows. Their immediate environment provides few suitable sites for egg-laying, and a nest keeps the eggs together and provides a convenient platform for the incubating parent.

Nest-building behaviour has been studied in a variety of birds and each species seems to have evolved its own technique. Blackbirds begin by collecting coarse twigs to make the base of the nest. They then collect finer material for the sides. When the framework is complete they collect mud to form the next bowl and line the bowl with hair and fine grass. Among canaries, the female does more of the nest-building. Her behaviour is stimulated by the hormone oestrogen produced by her ovaries.

Nest-building behaviour seems to be inherited. Evidence to support this comes from experiments carried out on lovebirds. Two closely related species of lovebirds show different patterns of nest-building

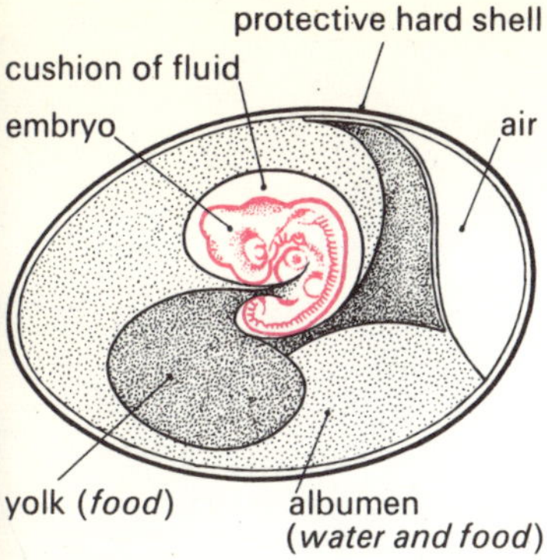

Note how alike the hen and human embryos are in the early stages. In the human, below, the tube-like brain, the limb buds, the eye and the vertebral column are all visible. It is seven weeks old, 12 mm long.

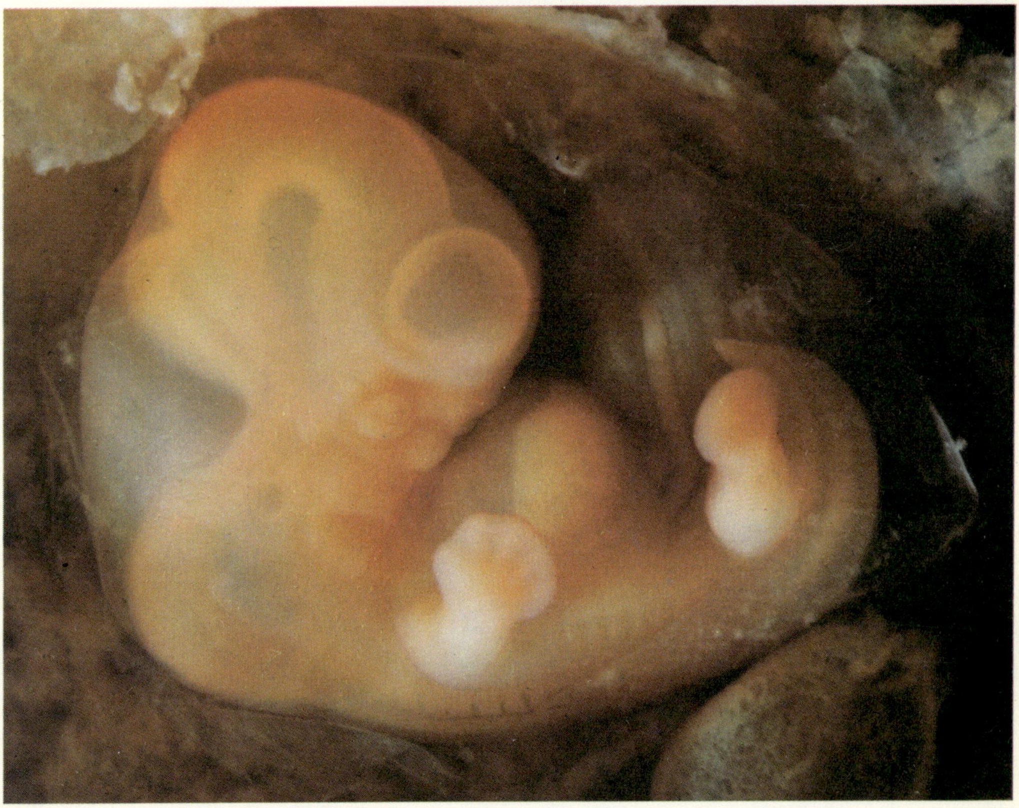

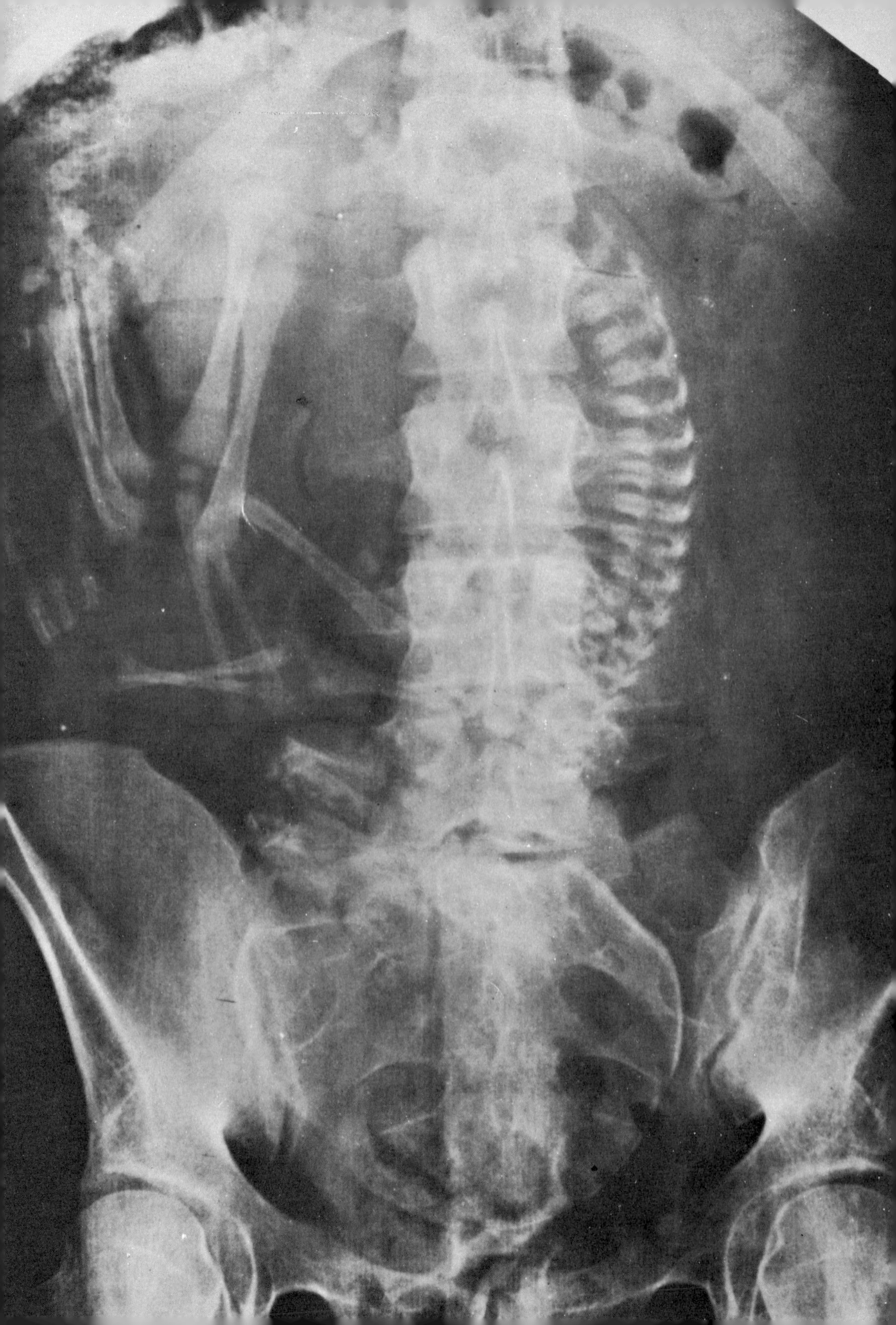

behaviour. One species, the peach-faced lovebird, tucks nesting material into the feathers above its tail. The other species, Fischer's lovebird, carries the materials in its beak. The two species were cross-bred and when the resulting young grew up, their nest-building behaviour was observed. These hybrid birds could not build nests for some time because they could not carry nesting material. They tried to combine the carrying methods of both their parents. They would try to carry nesting material in their beaks, but would drop it, or they would try to tuck the material between their body feathers but forget to let go. Lovebirds learn quickly, but even two years later the birds were failing in six out of ten attempts.

Food

Shelter is not the only advantage which young animals derive from the presence of their parents. They also have a much better chance of receiving a regular supply of food during their development. At birth or hatching, many animals lack the skill and physical strength to find their own food. A regular supply of food at this stage gives them time to mature.

The eggs of reptiles and birds contain a large supply of food in the form of *yolk*. In reptiles, this is the only food supplied to the young. When young reptiles hatch they are able to feed themselves. Hatching crocodiles snap and fight on their way to the water and immediately attempt to catch fish and other food.

In most kinds of birds the picture is rather different. Young birds may be born blind. Such chicks obviously cannot find food for themselves and must be fed by their parents. Feeding behaviour provides many examples of the stimulus-response pattern we introduced earlier. Young thrushes are blind. They cannot see a parent approaching with food. Instead, they respond to the vibrations caused by a parent bird landing on the edge of the nest. Their response is to stretch their necks straight up and open their beaks. Later, when their eyes are open, they will respond to the shadow of an approaching parent. However, at this stage, their response is still to stretch straight up. When their eyes have been open for a few days and they are stronger, they will *orientate* (direct) their response towards the

When a gall wasp lays its egg under the surface of the branches of an oak tree, the tree itself reacts by producing a round swelling or gall. This provides the larva with food until it is ready to emerge through a hole.

The human baby just before birth is X-rayed in its mother's uterus.

The changing responses of thrush nestlings.

approaching adult. This behaviour shows a changing pattern of stimulus and response as the young develop. Finally, the thrush's response becomes very accurate and is aimed towards the head of the parent—the point from which they receive their food.

Young gulls beg for food by pecking at the *bill* (beak) of the parent. Herring-gulls have a yellow bill with a red spot on it. Two scientists tried to find the stimulus which directed a young gull's pecking response. They made a large number of models of the head of a gull and offered these models to young gulls and noted their responses. It seemed that the young gulls responded to a sign-stimulus connected with the adult's bill, and that the shape of the head was not important.

Mammals provide food for their unborn young through a structure called a *placenta*. The placenta carries food and oxygen from the bloodstream of the mother to the young and carries away wastes from the developing embryo. After birth, a mother feeds her young for a while on milk produced by her *mammary glands*. As the young grow stronger she introduces other foods. Young *herbivores* (plant-eaters) start to look for plant food while still receiving food from their mother. By the time they are weaned (no longer receiving milk) they are able to find enough food to keep themselves alive.

Young *carnivores* (flesh-eaters) are fed by their parents even after they have been weaned. Wolves will kill their prey and swallow large pieces of flesh which they carry back to their dens. There they will *regurgitate* (bring back into their mouths) the meat for the young to eat.

Some animals make sure that they will have food for their young early in the breeding season, often even before they mate. They do so by establishing a *territory*, an area which they will defend against other members of the same species which would compete for exactly the same kind of food.

Protection

Parents protect their young from enemies which would otherwise kill and eat them. This protection is provided in many ways.

Homes and territories: homes provide shelter and territories safeguard food, but both help to protect the young against enemies, too.

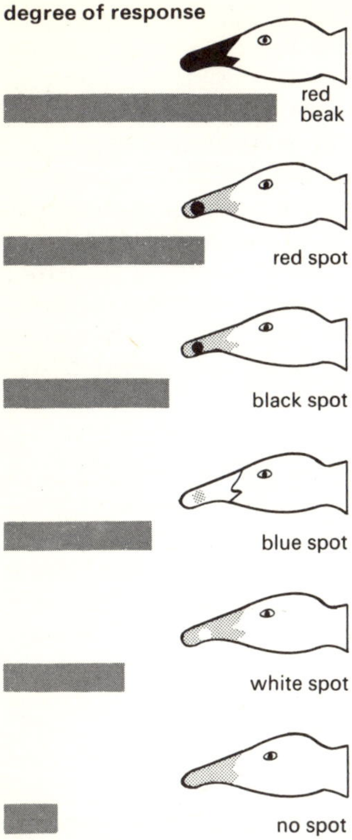

Spots attract gull chicks to peck the beak of an imitation gull's head, especially a red spot, as in nature. A beak coloured red all over attracts the best response of all, as the grey bars show.

Keeping the young together: if young animals are kept together it is usually far easier for the parents to protect them. In the jewel cichlid, a fish, both parents take care of the young. One parent stays on guard while the other feeds. They change roles at intervals. The young fish are kept together by a sign stimulus. This is the parent's zig-zag swimming movement which attracts the young fish. When one of the parents returns from feeding it swims straight into the middle of the shoal and begins to zig-zag. This attracts the young from the other parent, which swims off in a straight line in order to feed.

Another fish, the Egyptian mouthbreeder, carries eggs in its mouth until they hatch. Even when hatched, the fry will swim back to their parent's mouth for protection when danger threatens.

One type of learning behaviour appears to be adapted to the process of keeping young animals together. It is called *imprinting*. Some birds, such as pheasants, ducks, and so on, can walk as soon as they leave the egg. These birds learn to recognize their mother very soon after they hatch. If the mother is not present they will recognize as their 'mother' almost any moving object which catches their eye. As soon as this process of imprinting has occurred, the young will follow the mother closely wherever she goes. You can see that imprinting plays a very important part in keeping the young of some birds together.

Confusing the enemy: small songbirds are normally afraid of owls, stoats, and other enemies. However, in

Frightened by the photographer, the young Tilapia fish rush for shelter into the mouth of their mother.

the breeding season they become aggressive and will attack their enemies if their young are in danger. Enemies will often flee in the face of strong and surprising attacks by birds much smaller and weaker than themselves, because it is an unusual form of behaviour to which they have no instinctive defence.

Other birds confuse their enemies by pretending to be injured. If an arctic fox approaches a nestling snowy owl, one of the parent owls will trail one of its wings as if it is broken, and struggle away from the spot where its nestling is sitting. The fox, seeing the chance of an easy meal, will follow the 'injured' adult which will take to the air as soon as it has led the fox well away from its young.

Camouflage: the females of many ground-nesting birds have plumage of such a colour that it blends with the colour of the earth and makes them difficult to see. Eggs and chicks of these birds often have similar colouring (called *cryptic coloration*). When cryptically coloured animals sense danger, they *freeze* (remain completely still). This increases their chances of survival because many predators, particularly snakes, are sensitive to the slightest movement but may fail to see a perfectly immobile chick.

Hiding in this picture are two nightjars.

Carrying the young: when young kangaroos are born they are completely helpless. They crawl up through the hair of the mother's abdomen until they reach her pouch. Once in the pouch, the young animal remains there for protection. When it is strong enough to do so it will leave the pouch for short periods, returning every time danger threatens.

Monkeys also carry their young with them. Marmosets, small South American monkeys, usually give birth to twins. After birth, the female passes the young to the male, who carries them round clinging to his fur. The male hands them back to the female to feed. Even new-born human babies have a strong grip and can support their own weight by holding on with their hands. Could this ability have lasted from the days when our tree-living ancestors carried their young about with them?

Carrying has advantages for both parents and babies. The young are constantly protected and fed by

the parents and can be whisked away from danger. In addition, the parents can roam over a wide area in search of food.

Young baboons cling to the backs of their parents with four feet and tail when the troop is on the move.

Protection by breeding behaviour: one species of American robin shows a variable pattern of breeding behaviour. Some pairs breed during the wet season when earthworms, the main food of the nestlings, are abundant. However, the majority of these birds breed during the dry season, when earthworms are in short supply. The young robins hatched at this time have to be fed on fruit, a much less valuable food, and are constantly on the point of starvation.

This appears to be a most unreliable way of making sure of the survival of the species, but if we examine the behaviour of animals which feed on young robins the advantages of breeding in the dry season are easier to understand. In the wet season when food is in short supply, many enemies will seek out and destroy nestling birds. Of the young robins hatched at this time, only 15 per cent survive. However, in the early part of the dry season, many of their enemies have no food shortage and spend less time hunting for nestling birds. Thus, of the robins hatched at this apparently unsuitable time of year, about 40 per cent survive.

6 Growing and Learning

Parental care, which reaches such an advanced state in mammals, gives new-born animals a chance to grow up in a sheltered environment. Growing up means much more than just increasing in size. In the following pages we will be looking at different aspects of this process.

Physical growth

The most obvious aspect of growth is the development of the body. This begins as soon as the egg is fertilized. At some stage during this development, a young bird hatches or a mammal is born. Birds vary in the stage of development at which they hatch. Young pheasants and chickens hatch with a complete body covering of soft, downy feathers. Their eyes are open and, within a few hours, they are able to run about and start to feed themselves. Guinea-pigs are born with their eyes open and with a body covering of fur. They, too, can run about soon after birth. Compare these with young thrushes or human babies. Why is it so important for young pheasants and guinea-pigs to be able to run about so soon after birth?

Rates of development also vary. Young mice are fully grown and able to reproduce within a few weeks. Human beings are not able to reproduce until they are twelve or thirteen years old, and their bodies are not fully developed until several years later.

As you grow, your body increases in size, but it also changes in other ways. For example, the proportions of your body change. A baby has a comparatively large head and short legs. These proportions change gradually throughout childhood. During adolescence further dramatic changes occur under the influence of sex hormones. The great majority of babies are born normally and their development after that is normal also. However, problems do occur sometimes. Things can go wrong basically for two reasons. A baby may inherit defects from its parents or it may be harmed by something in its surroundings.

An example of inherited defect is *polydactyly*. Polydactic children develop six fingers on each hand and six toes on each foot. A simple operation removes the extra fingers and toes. Another physical defect which appears to be inherited is *mongolism*. A mongol child has a small, round head and unusual 'Asiatic' features. The child is usually small for its age. These physical defects are linked with other defects associated with mongol children, which we shall consider later.

Spastic children are often of high intelligence, but they have difficulty in co-ordinating their movements. This boy cannot speak, and communicates with his teacher by pressing the pads with his head.

Defects can also be caused by a child's environment. They may occur before a child is born. For example, German measles is normally a mild disease with few serious effects. However, if a women catches it during the first three months of pregnancy, the disease may have serious effects on the eyes and ears of her unborn child.

One of the venereal diseases, *syphilis*, can also seriously damage an unborn child. It attacks the embryo nervous system, and when the baby is born it may even be mentally defective. Syphilis can damage a baby's eyes and ears; it can also cause paralysis. For these reasons, men and women intending to marry should have a blood-test if there is any chance that they may have caught venereal disease at some time in their lives.

Even after birth there are still hazards in the environment which may prevent a child from developing normally. It may not get certain vitamins in its diet. For example, insufficient vitamin D affects growing bones and causes *rickets*. Lack of protein can cause *kwashiorkor*, a common disease in Africa, Asia and Latin America where many people live on a starchy diet.

Development may be faulty for a third reason; there may be some fault within the child's body itself. Hormones produced by the endocrine glands are most important to normal development. If a child has an underactive thyroid gland, for example, its development is seriously hampered. Such a child grows too slowly and may become very fat. It will be pot-bellied and appear lazy. Its intelligence will be low. This condition is called *cretinism*. If it is diagnosed in its early stages, injections of thyroxin can correct the balance and allow the child to develop normally.

These few examples serve to show that although in the majority of cases physical development is normal, it can be upset by influences from the child's parents, from its environment and from within its own body.

Starting to move about

As it grows, a young animal begins to move about. Children vary greatly in the rate at which they develop, but the sequence usually follows a similar pattern. In general, a child learns to control its muscles in sequence, starting with its eyes and working down through its body. So a child will learn to raise its head, then its shoulders, to roll over, sit upright, and crawl backwards or forwards. All this can happen before a child is seven months old, but it may not occur until much later.

At some stage after this an infant will pull itself upright and will eventually take its first steps. This is a complicated act. A young child has a large head and small feet, so balancing is difficult. Add to this the problem of lifting one of these small feet and balancing the body on the other foot in order to walk, and the procedure is even more complicated. To help overcome the problem, a baby often begins by pulling itself up alongside a suitable support and practises walking while holding on to objects along its path.

For the big-headed human baby, balancing on two legs is difficult.

The development of speech

The development of language begins in infancy and continues throughout life. Language can include gestures, but we are here concerned with the development of speech. The first sounds a baby makes are usually noises produced at random by its vocal organs. They have no meaning and are not produced in any pattern. It is worth noting that whatever the nationality of the baby, its early sounds are the same.

The next step in development is called *babbling*. A child will make a simple sound, for example, 'ma' and will then imitate this sound: '-ma-ma-ma'. This constant rehearsal of simple sounds will often continue for long periods. Through rehearsal the child learns to control its vocal organs. The child will then begin to imitate

As the pheasant chick grows, it learns to distinguish between harmless and dangerous silhouettes.

sounds made by other people if those sounds are particularly important to him. If a mother hears her child say 'ma-ma', she will probably make a fuss of him and repeat the sound. The baby will learn that saying 'ma-ma' produces a pleasant response from his mother and will repeat it whenever he wants attention. When he finds that words can have such pleasant results, he will begin to imitate other words in hope of reward. The child will speak his first word when he is about one year old, but this time varies widely. By the time a child is four, he may be using well over 500 words.

Opportunity to learn

You have already read that the behaviour of animals and man is part innate and part learned. Learning is a complicated process which continues throughout life but if an animal is to survive, there are some very important things it must learn while it is young.

The young of ground-nesting birds such as pheasants crouch when anything flies overhead. This freezing is an innate response and serves as protection against birds of prey. However, it is a wasteful response because it means that the pheasant chick crouches and freezes in response to the overhead flight of many harmless birds such as ducks, geese, and small songbirds. As the chick grows, it becomes accustomed to the overhead flight of common birds and learns to ignore them. However, birds of prey are relatively uncommon, so the typical long-tailed, short-necked silhouette of the predator and its pattern of flight remain unfamiliar and will still cause the pheasant chick to crouch. The process by which animals learn to disregard unimportant stimuli from their environment is called *habituation*.

Animals often learn by *imitation*. This is particularly true of children. You have already read that children learn by imitating their own actions in the development of speech. However, they also learn by imitating other children and adults as, for example, when a young girl plays with her dolls.

Reward and punishment play a large part in learning. Some actions produce rewards—a crying child may be picked up. Other actions produce punishments—touching a hot fire causes a painful burn. Reward and punishment are called *reinforcement*. The process by which young animals learn that an action will produce a particular reward or punishment is called *associative*

In an American laboratory classroom, a correct answer brings a smiling face to the pupil's television screen. Wrong answers get a sad face with a tear. An example of reinforcement used in teaching.

learning. Other types of learning are described by students of psychology and behaviour but go into rather more detail than we need.

How can we define learning? It is difficult to find an exact definition, but we can recognize that *learning has occurred if an animal's behaviour changes as a result of experience*. Notice that our definition does not tell us what learning is, but simply how we can recognize that it has happened.

Even if an animal or child is given every chance to learn, there are some things which it cannot do until it has reached a certain stage of development. Indeed, the amount of practice does not seem to be important. Even with little practice, the skills will still appear.

An example will make this clear. Two similar groups of tadpoles were used in an experiment. One group was placed in water and allowed to develop normally. The other group was placed in water containing an anaesthetic (*chloretone*) which allowed the tadpoles to develop normally but prevented them from moving. The tadpoles in ordinary water began to swim after spending several days in what seemed to be practice swimming actions. After the normal tadpoles began to swim, the drugged ones were placed in fresh water. After half an hour they were all swimming freely. It seems that the swimming response appears at a certain stage after hatching and that it does not depend on practice.

The process by which abilities appear at different stages of growth is called *maturation*, and it is a common feature in man and many other animals. Young children in our society are given freedom to move from the moment they are born. They can kick in their cots and prams, crawl as soon as they are able, and eventually learn to walk. No amount of coaching will make a child walk before it is sufficiently mature. Furthermore, once it is mature enough, it needs very little practice. One tribe of American Indians, the Hopi, carry their young children strapped to a board until they are nearly two years old. When they are released from the board, they walk easily with virtually no practice. Their bodies have matured to the necessary stage.

Increasing awareness

A new-born baby is very much an individual. It responds to its own needs—for warmth, food, and so on—rather than to information from its surroundings. It does not distinguish between human beings and the rest of its surroundings until it is two months old. At this age, the child learns to recognize human voices as something distinct from other sounds.

As a child grows, so its perception of its surroundings improves. Between the ages of three and four months, an infant can tell the difference between smiles and frowns, between kind and angry tones, and may also notice other children for the first time.

A young child also learns gradually to distinguish its own body from its surroundings and the bodies of other people. How can this come about? If a baby at play seizes its toes, how does this differ from the mother holding the child's toes? When the mother touches the baby's toes, the child's brain receives messages from the toes which are touched. However, when the child touches its own toes, it receives messages not only from the toes which are touched, but also from the fingers which are touching them. Similarly, when he makes noises, he can hear the sounds and also feel himself making them. In this way, the young child becomes aware of 'self' as something distinct from other people.

A two-month-old child responds in a simple way to other people. It will stop crying when picked up. Within another two months it will start to smile in response to smiles from its mother. From the age of nine months or a year a child will play happily, but this play is still very individual. Even when placed in a group, it will generally ignore the others. It is not until a child is about five years old that he will begin to play in a group, and it is at this stage that co-operative and competitive aspects of behaviour begin to develop.

Play

Play is an important factor in the development of many young mammals because it is in the play sessions with parents and litter-mates that they begin to learn the skills which will be necessary for their survival as adults. Lion cubs at play will stalk each other, crouch, and spring. These may be built-in parts of their hunting style, but they still have much to learn. As soon as they are able, they will follow their mother hunting and watch her killing her prey. Before they are a year old they will be skilful at stalking and approaching their prey upwind, but they will not make their first 'kill' until their mother suddenly leaves them when they are about two years old.

Children have far more to learn than the young of other animals before they can become fully independent. Young children begin to learn adult patterns of behaviour early in life and, for this reason, play is as important to them as it is to other animals. At play, a child rehearses for its entry into adult society by

Children use toy guns and bows and arrows in play. The South American Indian child, far left, will need to use his bow in adult life.

playing role after role. A young boy may be very happy playing with dolls. But adults, usually his parents, will lead him away from these to play with toy cars, modelling kits, guns, and so on. In this way, play helps to prepare children for distinct male and female roles. Do you think this is desirable?

Other influences on social development

All members of a family group are closely concerned with the development of children, particularly in the early years of their lives. However, most countries have developed schools which participate in the education of children. Schools teach a child about its environment. They also attempt to increase a child's usefulness as a member of society. In addition to its more obvious jobs, a school also provides a child with useful social education. A child at school gains experience of living and working with others. He learns to co-operate and to compete. Slowly, a child develops a pattern of behaviour acceptable to others and appropriate to his age.

The problems facing a growing child increase during adolescence. For example, teenagers have to learn about courtship. They must learn to conduct their courtship in a socially acceptable way. This is one example of the pressure put on young people in an effort to make them conform.

Adolescence is also the time when people progress from a passive to an active role in society. As children, they have accepted their low-ranking social position and have usually obeyed the instructions of both parents and other full members of adult society. Adolescents begin to challenge the authority of their elders.

Emotional needs also require satisfaction. Young people need to be able to respect someone and this may develop into hero-worship of a current idol. They also need the respect of others. This need can be satisfied in several ways. In teenage gangs, members may fight for control and a definite order of rank may be established. Some adolescents may seek the respect of others by acts of bravado. Unfortunately, such acts often make the individual look foolish, and so can diminish their standing in other people's eyes. Often, all that is required to gain the respect of other people is to show an increasingly mature attitude to life.

This is a very cosy view, of course. It is equally

possible that an adolescent will develop aims and ideals opposed to those of the society in which he lives. This can, and does, cause serious conflict. Conforming with the ideals of the society in which you live produces harmony and reduces friction, but it is only through rebellion that society can advance. Rebellion need not be violent, but it is a mistake to accept without protest any social ideal which you consider to be wrong. Remember that childhood is basically a selfish stage, whereas adolescence is often marked by the development of a strong feeling for social rights and wrongs. Every individual has the power to contribute something to a society.

The effect of heredity and environment on development

So far, we have been looking at the development of children and other animals in general terms, but it is obvious that animals are different. Not only are sheep different from goats, but one sheep is different from another. A shepherd can recognize individual members of his flock. It is also apparent that people are different;

The clothes and life style of teenage groups stand out in contrast with those of their elders. This is one of the aims of adolescent rebellion.

not only people of different races, but also people of the same age from the same area, and so on. Not only do they look different; they react differently in the same situation. What makes people react in different ways?

There are two main groups of factors which influence the development of individual patterns of behaviour. Some factors are inherited from the parents and others are found in a child's environment. This is also true of other animals. If a male cichlid fish is raised in isolation it will, when mature, go through its courtship routine when presented with a very rough model of a female cichlid. However, males that are raised in the company of other males and females will court only the living females or very accurate models. The male cichlid inherits the pattern of courtship behaviour, together with a rough idea of the shape of the female. However, the presence of females and other males of the same species gives the young male a chance to learn the exact shape of the female, and he will not be deceived by a rough model.

Another example of behaviour which is part-learned, part-inherited is found in polecats. Young polecats were reared in isolation and never given live prey during their development. When they were mature, each was given a live rat. Each polecat showed interest but made no move to attack the rat unless it ran away. As soon as the rat ran, the polecat pursued it, seized it, and shook it—a typical polecat reaction—but its first attempts to kill the rat were clumsy. However, after a few attempts, the polecat learned to seize the rat by the neck and kill it with a single bite. In this example, an innate behaviour pattern was improved by practice. When polecats are allowed to develop normally in the company of their parents and litter mates, they develop the necessary hunting skills during periods of play.

It is less easy to determine how much of a child's behaviour is inherited and how much is caused by the effect of the environment because it is impossible to experiment in this way on young children. However, it is possible to say that some children appear to inherit the tendency to be very active, while some seem to be 'born idle' (in terms of physical activity, at least).

The effect of heredity and environment on intelligence

Two main methods have been used to study the effect of heredity on intelligence. Sometimes, following a

Identical twins have a similar start in life. If they are brought up in contrasting environments, one twin may show more intelligence than the other.

tragedy like the death of the mother, identical twins have had to be separated shortly after birth and brought up in different homes. Identical twins inherit exactly the same factors from their parents. If intelligence were completely controlled by heredity, the intelligence quotients (IQs) of two identical twins would remain exactly the same whether they were raised together or apart. In fact, this has not been the case. When tested, identical twins reared together showed almost identical IQs, as did those reared apart but in similar surroundings. However, identical twins reared in very different environments showed IQs which differed by as much as 24 points. It appears, therefore, that environment must have a considerable effect on intelligence.

This is not the whole story. Similar tests have been carried out on *fraternal* (non-identical) twins. Fraternal twins reared in very different environments showed greater differences in intelligence than identical twins who received similar treatment. What conclusion would you draw from this result?

Investigations show that the differences in intelligence between parents and their children are smaller than those between foster-parents and their foster-children, even when these foster-children had been adopted as babies.

We can summarize this section by saying that heredity appears to set limits on a child's intelligence, but within these inherited limits the environment can have a considerable effect on the actual intelligence-level achieved.

The development of personality

Think of someone you know. Now write down the answers to the following questions: What does he/she look like? What is he/she really like—as a person? Now consider your answers. They are describing the same person but in completely different ways. In one case you are describing physical characteristics; in the second you are describing a *personality*. The total personality of an individual is made up of many parts.

We have already discussed the development of some of these parts, which we call *personality traits*. For example, intelligence is a personality trait. So is physique; there are many more.

Some of these traits are easy to see and are often called *surface traits*. These include the general level of physical activity, how friendly and sociable a person is, and so on. Other traits, often called the *inner personality*, are not easy to discover. They include fears, hopes, and desires of which we may be aware, but about which we do not talk very easily. There are also other aspects of our personality of which even we are not aware, and these traits can be discovered only by deep investigation. Even then it is not easy.

So, in general terms, your personality is made up of all those aspects of your behaviour which make you different from other people. Personality traits are often sub-divided into two groups. These groups are called *temperament* and *character*. Temperament is said to be inherited, whereas character is said to be the result of all the individual's experiences throughout life.

Effects on the development of personality

A child may inherit a high level of physical activity and a low level of intelligence. These two inherited traits will have a great effect on the sort of personality he will develop. In addition, other physical factors have an effect on personality. For example, a beautiful girl or a strong, handsome boy will have a good chance of being accepted by his age group, whereas a child who suffers from some physical deformity may become shy and withdrawn. (This is one of the reasons why you should not judge people on first impressions. It may be easy to like a handsome or beautiful person, but this person could also be emotionally deformed—selfish, dishonest or cowardly.)

The group in which we live also affects the kind of personality we become because each *culture* has certain attitudes and standards which are thought to be important. In European countries young people are encouraged to become independent. If they ask for help, it may be taken as a sign of weakness. In some African and Asian cultures on the other hand, independence is thought to be less important.

Similarly, we live in a very competitive society. Everyone is encouraged to look out for himself and one of the major motives in our lives is physical gain. Contrast this with an Israeli *kibbutz* (farm) or one of the increasing number of small *communes* growing up in various European countries, where co-operation and property sharing are the rule. Can you find out more about different culture groups? Which seem to be the most worthwhile features of these cultures? What features of our present culture would you most like to change?

In addition to the general effects of the culture in which we live, the *experiences* of a child growing up within that group will also have a considerable effect on his personality.

Just as environment affects intelligence, so it can affect other personality traits. If a child has a warm, close relationship with his mother, he will learn to trust her, and this trust will be extended to others. On the other hand, a child reared in a cold, unloving environment may become quiet, unhappy, and inactive. In extreme cases he may become emotionally disturbed. An emotionally disturbed child may act in a way which is not acceptable to the group. His behaviour may be considered unusual—abnormal even, and this leads us to consider what is normal and abnormal behaviour.

7 Abnormal Behaviour

Your behaviour is made up of all the activities you carry out. Physical behaviour consists of the physical activities of the body—those you control, such as movements of limbs and those you do not consciously control, such as the actions of glands. Mental behaviour consists of the activities of the mind—thinking, reasoning, remembering, and feeling emotions. This chapter is about *abnormal behaviour* and we must first of all think about what this means.

If someone you knew put a metal ring through his nose you would probably think he was behaving abnormally. Yet a woman who puts a ring of metal through her ear is considered to be quite normal! There is not really much difference between a ring through the nose and a ring through the ear, so why is one action thought to be abnormal?

Let us look at another example. If your neighbours went to live in a cave you would probably think that they were behaving abnormally. Yet thousands of years ago a cave was the normal place to live in. It is obvious that much so-called abnormal behaviour is simply *unusual* behaviour which does not cause anyone any harm.

So it must be stressed that you do not have to act in exactly the same way as your neighbours for your behaviour to be considered normal. It is very difficult to define normal behaviour for the following reasons.

Do not confuse abnormality with the unusual. If everyone practised yoga there would be nothing strange about this man's position. Yoga trains the mind and body—it is unusual but not abnormal.

1 All the factors which we have been examining combine to produce an individual who is unique. There is no one else exactly the same as you. Yet the

Hogarth showed 'Bedlam' in 1763 as a place of entertainment at the expense of the mentally sick.

behaviour of the vast majority of the very different individuals who live in your community is considered to be normal. For example, at a party, a person with slightly extrovert tendencies will move about easily and confidently, talking to anyone with little or no hesitation. A slightly introvert person may be shy in such a situation. These two individuals react in very different ways, yet both types of responses fall within the very wide limits of what is accepted as normal behaviour.

2 Behaviour which is considered normal at one stage may be considered abnormal at a later stage. Crawling is normal behaviour for a one-year-old child but would not be considered normal for a healthy 15-year-old.

3 Behaving normally is not the same thing as being well-behaved. When a very young child breaks up a toy the parents may not consider this to be an example of good behaviour, but it is perfectly normal. However, similar destruction by a 15-year-old would be considered abnormal.

In this chapter we shall be looking at abnormal physical and mental behaviour caused by illnesses and handicaps. These cause much unhappiness and often prevent a person from working, marrying, and so on. It is worthwhile to consider for a moment how we regard people who behave abnormally.

Less than two hundred years ago people would visit lunatic asylums to laugh at the insane. Today this seems a cruel thing to do, but are we really so much better? How many times have you made or heard jokes about 'loonies' or spastics? Some people react to abnormal behaviour by ignoring it and by trying to pretend that it is not happening. They do this because such behaviour frightens them or upsets them and they cannot face up to it. Some people are ashamed of abnormal behaviour and many people are too embarrassed to talk about relatives who are mentally ill.

Most of us would agree that it is wrong to laugh at or ignore the ill and the handicapped: what can we do to overcome these attitudes? Our attitudes are caused largely by ignorance and so we must try to find out as much as possible about the causes of abnormal behaviour. We must also take responsibility for the mentally ill by looking after them and making them feel that they are a part of our society and accepted in it.

Inherited abnormal behaviour

Scientists have discovered that there are several causes of abnormal behaviour which are inherited. One rare inherited disease, called *dystrophia myotonica*, affects one in 50 000 babies. The symptoms include wasting of the muscles and, in some cases, mental defects. Although the disease is inherited, it does not usually appear until the person is about 20 years old. The parents of a person with this disease may be perfectly normal themselves.

Another inherited disease is *mongolism*. The mongol child has a small round head, unusual features and is short for his/her age. Mongols are very unintelligent. They do not usually have IQs of more than 50, compared with the IQ of a normal person which is about 100. So mongols cannot do arithmetic although they can sometimes read and write. They often have good memories and are usually cheerful and friendly people.

The mongol child enjoys a ride on the merry-go-round.

Acquired abnormal behaviour

Once a person is born there are, in his environment, things which may cause abnormal mental and physical behaviour. There are diseases which can affect his behaviour temporarily or permanently. Accidents may damage the body and handicap normal behaviour. Drugs may cause abnormal behaviour and stresses may cause mental illnesses. It would take several books to describe all the things that can go wrong and so we will look at just a few examples.

As you read in Chapter 1, we use our senses to receive information about our environment. If our senses are defective, we do not receive all the information we need to behave normally. This point is obvious when you think of how a blind person finds his way around a room. The sense of sight can be affected in several ways. Something may go wrong with the eyes themselves—perhaps a *cataract* may form in the lens. As a result, the lens loses its transparency, or the nerve from the eye to the brain may become diseased. Blindness may be caused by damage to the brain which prevents it from responding to the impulses which reach it from the eyes.

The information we receive from the environment is used to make responses. Correct responses can be made only if the nervous system and muscles are working normally. *Locomotor ataxia* is an example of a disease which affects responses. This is a disease of the spinal cord and it can be caused by syphilis. People suffering from the disease lose their reflexes, walk unsteadily, and end up helpless and bedridden.

Drug addiction Today a large social problem, especially among young people, is drug addiction. Doctors use drugs for various reasons—to produce sleep, to calm nerves or to stop pain. Some drugs, such as heroin, can be bought only with a doctor's prescription. Other drugs, such as codein and alcohol, can be bought without a prescription and are used by large numbers of people. Cannabis is a drug which is not used by doctors at all and can only be bought illegally.

A drug addict is a person who has got into the habit of taking a drug and can no longer do without it. Some people think of drug addicts as criminals, whereas others see them as people who are physically or mentally ill. It is true that a person who buys and uses illegal drugs is committing a criminal act but we ought to look for reasons why drug addicts do these things.

Many young people start to take 'soft' drugs such as cannabis because they see their friends doing it. Some find that the drug relaxes them and makes them feel

'Kerstin, born 1954, died 1968. Narcotics took her life.' Swedish publicity against drug addiction.

Kerstin
*20/1 1954
† 24/12 1968
Narkotikan tog hennes liv.

good. Some of these find that they can 'take it or leave it'; in other words, they do not become addicted to it. Still others use the drug to escape from reality because they are unhappy or not adjusted to the real world. It is these people who can come to depend on the drug because it seems to solve their problems. Of course, it does not really solve any problems. In fact, it creates problems. If a person uses cannabis over a long period, he may become lethargic and unable to work normally. Another danger is that a person who is feeling relaxed under the influence of cannabis may be tempted to try more dangerous drugs. In recent years there has been a lot of argument about whether or not cannabis should be banned. Some people say that it is no worse than alcohol, but we will look at the problems caused by that drug a little later on.

Morphine is a hard drug which has much more serious effects than a soft drug such as cannabis. Morphine has similar effects to those of heroin. Both drugs are obtained from the unripe seeds of a kind of poppy which grows in Central America, Turkey, India, China and the area known as the Golden Triangle (where the borders of Burma, Thailand and Laos meet). Doctors use morphine because it kills pain, makes a patient cheerful, and sends him to sleep. Large doses of morphine can cause addiction, so doctors use it very carefully and give only small doses. To use morphine for the pleasant effects it produces is extremely dangerous. When the pleasant feeling wears off it is replaced by very unpleasant reactions known as *withdrawal symptoms*. These symptoms include dizziness, sickness, perspiration, diarrhoea, depression and anxiety. An addict will often take more morphine to get rid of the withdrawal symptoms. Unfortunately, the body comes to need more and more morphine to produce the pleasant effects. In this way an addict becomes dependent on larger and larger doses of the drug. The person on hard drugs is unable to lead much of normal life because he spends so much time and money trying to obtain the drugs he needs.

Alcoholism The National Council on Alcoholism has estimated that, in 1971, there were 520 000 alcoholics among Britain's working population. In terms of numbers then, alcohol is a much more serious problem than heroin is. Some people drink alcohol because it helps them to relax in company. Others drink it because it helps them to forget their problems. The effects of alcohol on behaviour are well known and fortunately most people can stop drinking at any time they want to.

Alcoholics are different, however. They are people who have come to depend upon alcohol to such an extent that they cannot stop drinking it. Alcohol taken over a long period of time poisons the nervous system and causes delirium tremens (DT). When a person is suffering DT he is restless, confused, and feverish; he may imagine fantastic things and claim, for example, that he is being chased by demons.

Many people despise alcoholics because they believe that alcoholics are responsible for their own troubles. For this reason, alcoholics are often ashamed of themselves and try to keep their addiction secret. In fact, nobody sets out to become an alcoholic and perhaps it is time for us to think of alcoholism more as a disease which requires treatment and sympathy like any other disease. Alcoholism has very serious effects upon industry. It prevents many people from doing their work properly and is also the cause of many accidents in industry and on the roads.

Alcoholics, alone in the city street.

Mental illness

About half our hospital beds are filled by people who are mentally ill or mentally deficient. The National Association for Mental Health has said that mental illness causes more loss of production than strikes or 'go slows'. In 1970, more than 36 million working days were lost because of mental illness. However, it is very difficult to assess the full effects of mental illness on the population because so many kinds of illnesses are not recognized even today. Even a person who is suffering may not know that he is ill, and so his illness goes untreated. This lack of diagnosis is serious both for the individual and for the society in which he lives.

What causes mental illness? This is a very complex question because the causes of mental illness are many and varied. Very often the cause lies not only with the patient but also with the society within which he is trying to live. We live in a very complicated world and the pressures on us are great. Some people fail to adapt themselves to this society and the strain of trying to do so can make them mentally ill. Strain can arise in many ways. In adults, it may be caused by trying to hold down a job which is too difficult, or by an unhappy marriage. Children may show symptoms of mental illness as a result of an unhappy home background.

Is the city environment good for the mental health of man?

There are many symptoms of mental illness. These may include excessive touchiness, anger, nervousness, and depression. Let us look at some kinds of abnormal behaviour arising from various types of mental illness.

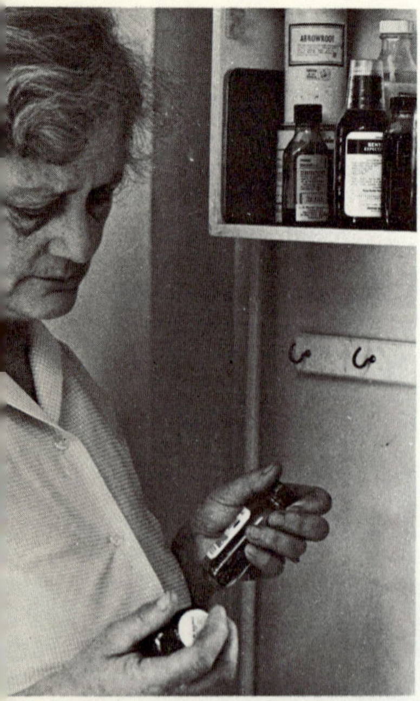

Hypochondria is a form of neurosis.

Neurosis is a relatively mild mental disorder. Neurotic people often go to the doctor and complain of various body illnesses although, in fact, there is nothing wrong with their bodies. These people cannot face up to the problems of life and so they use their 'sickness' as an excuse for not being able to cope. This is not a deliberate act but an unconscious method of finding sympathy.

Several types of neurosis are recognized:

Neurasthenia sufferers complain of a variety of illnesses, although doctors can find no bodily fault. Symptoms may include sweating and trembling, feeling hot and cold, weakness, and pains in various parts of the body.

These symptoms often appear in people who have no positive aim in life. Their illnesses, real or imagined, are the only things they have to think about. They also serve to take the person's mind off his failure in other fields.

Dissociation. This group of disorders is often given the collective name of *hysteria*. Some hysterical people become temporarily blind or deaf, or lose their skin sensitivity. Others develop twitching muscles or paralysis. There are memory disorders too, such as *amnesia* or *fugue*. Amnesia is loss of memory; people are sometimes found wandering about without any idea of who they are or where they have come from. A fugue is a state of mind where a person may carry out an act—even murder—and not be able to remember it afterwards.

Anxiety neurosis Neurotics are sometimes very anxious people. The anxiety may be about such things as enclosed spaces (this is called *claustrophobia*) or it may be about nothing in particular. These abnormal anxieties are caused by conflicts within a person. For instance, a homosexual might experience a conflict between his sexual urges and his fear of other people's opinion.

Neurotic people are not usually dangerous and do not have to enter mental hospitals. The various types of treatment used all concentrate on finding the causes of his conflict and encouraging the neurotic to deal with it realistically.

Psychosis is a much more serious mental illness than any neurosis. People suffering from *psychoses* (plural of psychosis) are known as *psychotics*. They are often out of touch with reality; they 'see' and 'hear' things which are not there and may imagine that they are someone else—often a powerful figure like Christ or Napoleon. Psychotics are often dangerous and have to be looked after in a hospital. In England and Wales, psychotics account for about half the number of mentally ill persons admitted to hospital.

There are two types of causes of psychoses: there may be something wrong with the body or there may be something wrong with the way in which the person has adjusted himself to life. It is quite possible that scientists may find out that a person cannot adjust to life because of some fault in his body, but for the present this is only a guess.

Syphilis is a well known example of a disease of the body which can cause psychoses. Syphilis can affect the brain and cause *delusions of grandeur*, as when a patient thinks he is somebody famous. Psychoses can also be caused by brain deterioration in old age and by the excessive use of alcohol.

Two types of psychosis which are not thought to be caused by faults in the body are *manic-depressive psychosis* and *schizophrenia*.

Manic-depressives suffer alternate moods of great happiness and depression. When a manic-depressive is in an 'up' mood he often imagines things and becomes so excited that he has to be restrained. In a 'down' mood he becomes deeply depressed and may even try to commit suicide.

Schizophrenia sufferers withdraw from normal contacts with other people. In one form of schizophrenia the patient sits and stares into space and appears to have no contact with the world around him. Another type of schizophrenic acts in a silly manner and talks nonsense. Another type sits or stands in an unusual posture for many hours. In yet another form of schizophrenia, the patient appears intelligent, speaks quite clearly, but talks about totally unreal events. He may imagine that all the people around him are spies, or that someone is trying to kill him, or even that he is already dead.

Psychopathy is a third category of mental abnormality and is a term applied to people whose personality development has been faulty. Psychopaths have no conscience or concern for others. They may lie or steal without hesitation. Some can even commit sexual crimes or do violence with no feelings of guilt.

The paintings of schizophrenics often show quite clearly their emotions.

Treatment of mental illness

The human brain is a very complex organ and we understand it poorly. For this reason treatment of mental illness is a tremendously difficult job, and this branch of medicine is still not as well developed as are many other fields. However, a variety of methods of treatment has been developed and they are used with considerable success today. Let us look briefly at some of the more common forms of treatment of mental illness.

Group therapy is a form of treatment in which several people meet under the guidance of a psychiatrist to discuss their problems. The advantages of group therapy are that it is often helpful to an anxious person to discover that other people have problems similar to his own, and that group members can help each other in their efforts to find solutions to their problems. Can you think of any well known organization which works in this way?

Shock therapy is used in two ways. In *electric shock therapy*, a small electric current is passed through the patient's brain, producing a mild convulsion followed by temporary unconsciousness. In *insulin shock therapy*, insulin is injected into the bloodstream, causing the patient to fall into a coma. After a short while, doctors raise the concentration of sugar in the blood to the normal level by injections and the patient recovers consciousness. Both types of shock treatment

Working together at physical tasks is a form of therapy and helps to give the mentally ill a constructive role in society. Electric shock therapy, right, is given under anaesthetic.

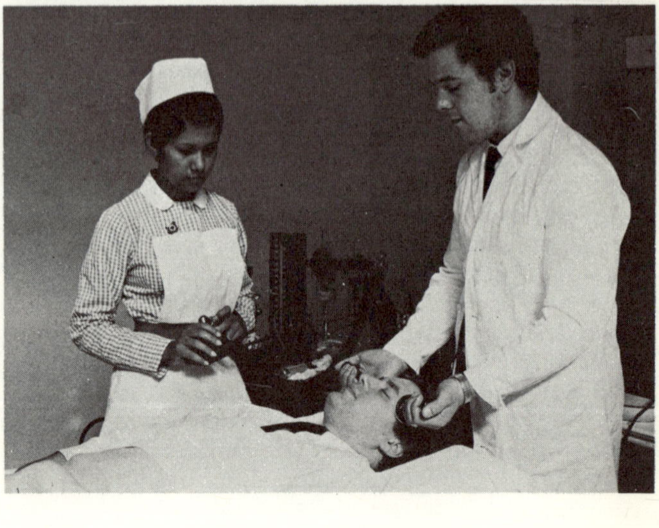

have been used with some success in the treatment of manic-depressive and schizophrenic psychoses.

Drug therapy Many drugs have been investigated in efforts to bring mental patients back to full health. Tranquillizers are used regularly to treat schizophrenics, for example. One of these, chlorpromazine, appears to have a degree of success higher than either of the forms of shock therapy.

Psychoanalysis usually consists of the classical procedure of persuading the patient to relax in an armchair or on a couch and talk freely about anything which occurs to him. The analyst will guide his thoughts by asking occasional questions. Successive meetings lead to deeper and deeper questions and answers as the analyst probes for the problems of his patient. Some interviews may become very emotional, but can lead to a release of tension and help both analyst and patient to understand the deep-rooted problems causing the patient's anxiety and so suggest appropriate action. The analysis of dreams may also figure as one part of psychoanalysis.

Brain surgery may be used in certain cases of mental illness. Prefrontal lobotomy is an operation in which the surgeon destroys the frontal lobes of the brain. This operation has improved some schizophrenics and manic-depressive psychotics but it is not used much now as it has largely been replaced by other more satisfactory treatments.

Avoiding mental problems

What can you do to keep yourself in good mental health?

1 Develop the habit of talking about your worries and problems with a close friend. (In the words of a well-known proverb: 'A trouble shared is a trouble halved.')

2 Take care of your physical health; worrying about physical illness can often cause mental stress.

3 Try to overcome problems you encounter. Don't sit and worry about them. Develop contacts with other people. Try to find a job which interests you. Use leisure time to relax.

In other words keep fit, keep problems in perspective, and try to enjoy life to the full.

8 Animals in Groups

Space is at a premium on the Bass rock, where gannets nest in their thousands. Each pair has a minute territory which it will fiercely defend against intruders.

In the previous chapter we looked in some detail at patterns of development in man and other animals. Although things do occasionally go wrong, the majority of us develop in a perfectly 'normal' way. Our pattern of development leads us gently from our family group and into a larger social community. We will now consider the most important features of behaviour in social groups in man and other animals.

Many animals live in social groups. You have already learned something about two specialized kinds of groups: the pair and the parent/offspring group. Animal groups can take many other forms. For example, moths will gather round a light. They are all guided to the same spot by a directional stimulus, but that is all they have in common. They do not form a social group because animals in a social group react to the behaviour (signals) of other animals of the same species. The moths respond to the light, but not to each other.

Notice that a social group can consist of as few as two animals. It may also consist of thousands of members, as it would in a colony of termites.

This definition of a social group implies that there is some form of communication between social animals. How many examples of communication can you remember from previous chapters? Can you remember the signals and responses we mentioned in communication? Here are some reminders.

Male fireflies look for *flashing light* signals to locate females. Young gulls peck at a *colour* (red) on the adult's beak. They also respond to a *pattern*—the contrast between the spot and the general bill colour. Young thrushes respond to a *mechanical stimulus*—the vibration of the nest as the adult lands. Female crickets move toward the *sound* of a male. Male gipsy moths respond to a *chemical* stimulus or pheromone.

Light, colour, pattern, mechanical stimuli, sound, pheromones—you have already met a wide variety of signals used in communication. Communication is the key to the success of any society of animals. Look out for other examples as we explore larger social groups.

Animal groups can be divided into two broad categories: insect societies and vertebrate groups.

Insect societies

Social insects include ants, bees, wasps and termites. These groups consist of a large number of individuals, often descended from one female. Within the society there is a *division of labour*: each individual has a role which it fulfils instinctively. Ants and bees have been much studied. An ant colony may begin on a warm still day in summer, when winged males and females emerge from neighbouring nests and take to the air. This flight is often called the *nuptial flight* or *wedding flight* because, as they fly, the males and females mate. Sometimes, the nuptial flight is very short and mating may occur on the ground. This is the only time in her life that the female will mate. She takes many thousands of sperm from her mate and stores them for later use. The only function of the male is to fertilize the female and when this is done the males soon die.

Immediately after mating, the female sheds her wings. She will never fly again. Her behaviour after this varies. She may return to her own nest or to another nest of the same species, or she may start her own nest. To do so, she finds a small hole in which she can wait. She lays eggs and, when they hatch, she feeds the grubs (which are completely helpless) on food stores from her own body until they *pupate*. When the *pupae* have developed, she opens them to allow the new adults to emerge.

These adults are immature females. They are sterile and cannot breed. Sterile females are called *workers*. They immediately take over the job of maintaining the colony. Some of the workers care for the nest and look after the young. They move the young near the surface of the nest on warm days and deeper in to the nest when the weather is cold. They feed the grubs and dig

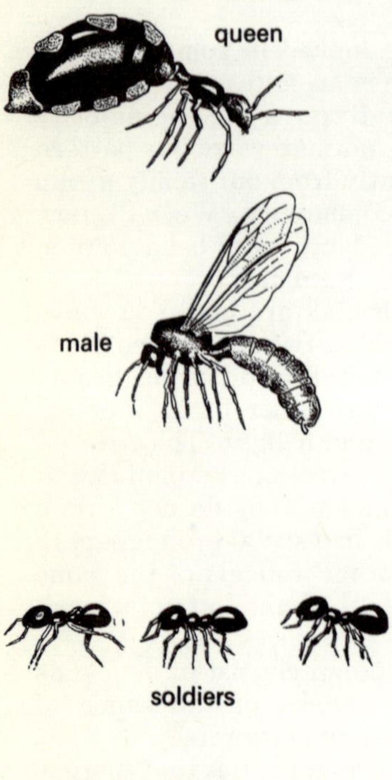

Different members of an ant society.

Ant larvae are fed and tended by the workers, which carry them about regardless of effort so that they are always in the optimum conditions and temperatures.

new tunnels, and they open the pupae to allow young adults to emerge. Meanwhile, other workers go to forage for food.

Food is the key which binds the colony together. Foraging workers carry food back to the nest and give it to the other workers who have remained at home. Hungry workers beg for food from those who have eaten and so the food is passed from one to another. This exchange of food is called *trophallaxis*. It ensures co-operation between ants of the same colony and appears to give them all a common scent. This means that intruders in the nest can be recognized quickly.

When ants encounter another ant or possible food, they touch it with their *antennae* (feelers). In the antennae are sense cells similar to our receptors for smell and taste. Because ants spend much of their lives in darkness, this sense is very useful, and ants make use of smell and taste 'landmarks' in the nest. It also helps ants on their foraging trips outside the nest.

Ants seeking food wander widely from the nest. When they find food, they carry it back to the nest quickly. As an ant returns to the nest with food, it touches the ground at intervals with the tip of its abdomen. As it does so, it leaves a drop of liquid on the ground. This liquid is a pheromone. Ants leaving the nest touch the ground with their antennae and use the 'scent trail' to guide them to the food supply. As these ants return to the nest they, in turn, leave more pheromone to reinforce the trail. As the food supply decreases, the ants returning to the nest will not be very excited. As a result, they will not reinforce the trail and it will fade away. (If the trail persisted for a long time the nest would be surrounded by hundreds

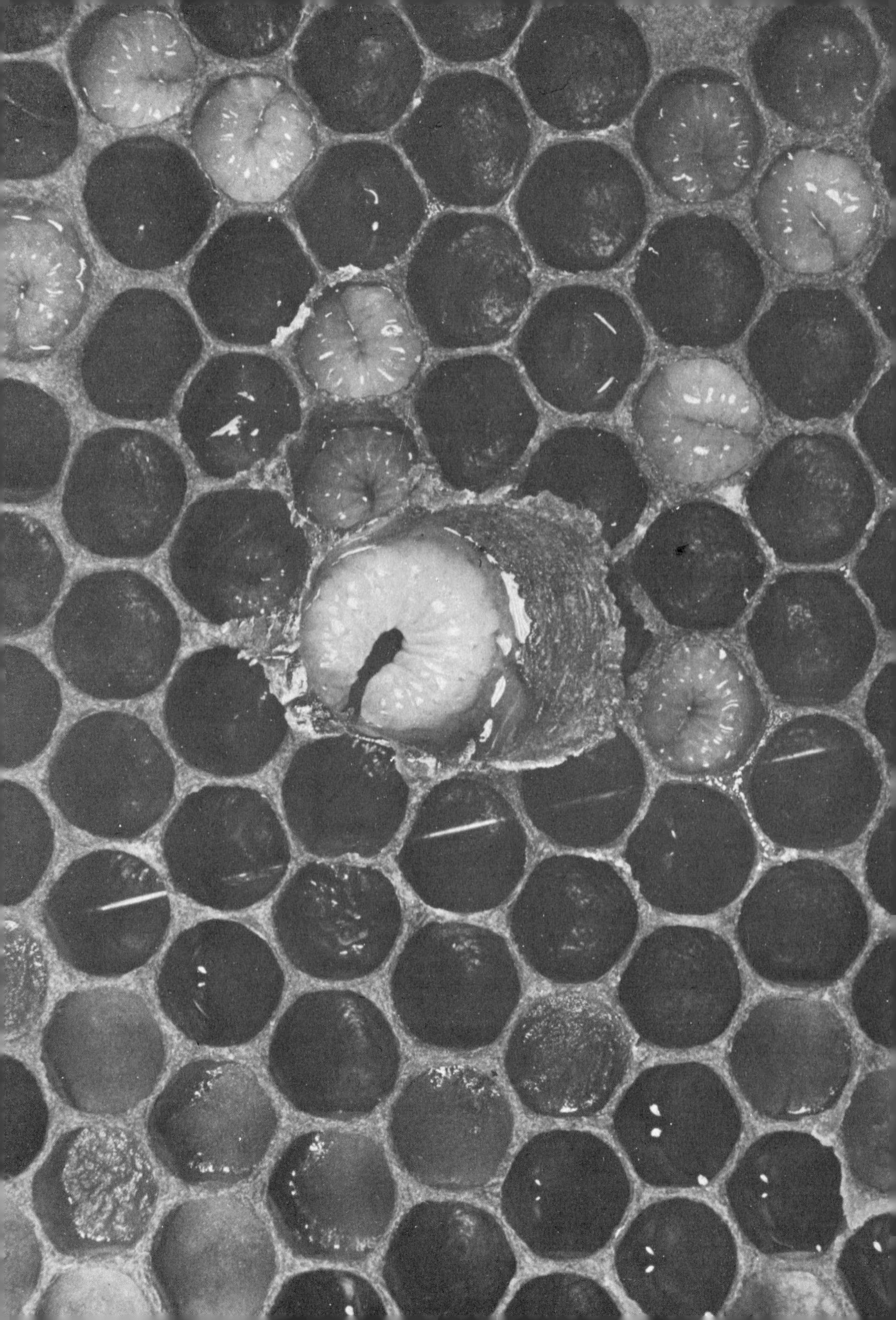

of trails, and very few of them would lead to food supplies.) Experiments with one North American species, the fire ant, suggest that a 20 cm trail will last less than two minutes if it is not reinforced, whereas trails laid by the army ants of Central America may last for up to a month in the dry season.

The social organization of bees is similar to that of the ants in many ways. The colony usually includes one fertile female, the *queen*, but the majority of adults are once again workers (sterile females) who build and maintain the nest and care for the young. As in ant societies, there is division of labour, trophallaxis and communication in connection with foraging for food.

Division of labour among worker bees is related to the age of the individual bee. During its life one bee will carry out different tasks within the society. A newly emerged female works in the hive. When she is a few days old, she starts to feed the grubs and, later still, helps to build new cells in the nest. When she is about two weeks old she may take up guard duties at the entrance to the hive, or she may go directly to foraging duty.

Bees on guard duty check every insect landing at the entrance to the nest. Members of the same colony have the same colony odour and will be allowed to pass. Strangers will be seized. If they fight back, these strangers will be killed, but if they submit to inspection they will be allowed to fly away unharmed.

Left: the large larva, nourished with a special substance will become a queen bee. Above: guards check a bee for odour. Right: they kill an intruding wasp.

Mouth to mouth exchange of food between two bees. Right: a bee returning from a distant food source dances a waggle dance on a vertical surface of the hive. The direction of the dance will be upwards if the food is in line with the sun. The bee will dance at an angle to the vertical corresponding exactly to the angle between the food source and the sun from the hive. Attendant bees will find the food.

Trophallaxis in bees is similar to that in ants. Hungry bees will beg from bees with food and will be rewarded. However, food from foraging is not the only substance passed on from worker to worker. The queen bee produces a chemical often called 'queen substance'. The worker bees who take care of the queen pick up this substance from the surface of her body and pass it to other workers by trophallaxis. As long as the queen substance circulates, all cells built by the workers will be normal, small hexagonal cells. But if the queen dies, or the flow of queen substance is interrupted for any other reason, the workers will build large *royal cells*. Grubs in these cells are specially fed and will develop into replacement queens. This response ensures the survival of the colony.

Communication between foraging honey-bees does not depend on scent trails, but on a pattern of behaviour called *dancing*. Bees which have found a rich food supply perform a dance when they return to the hive which tells the other foragers not only the direction of the food but also how far it is from the nest. When food is less than 90 m from the nest the returning forager will dance a round dance. This dance stimulates the other workers in the hive to leave and forage for the food. They are guided to it by the scent of the dancing bee's body and also by food which they receive during the dance.

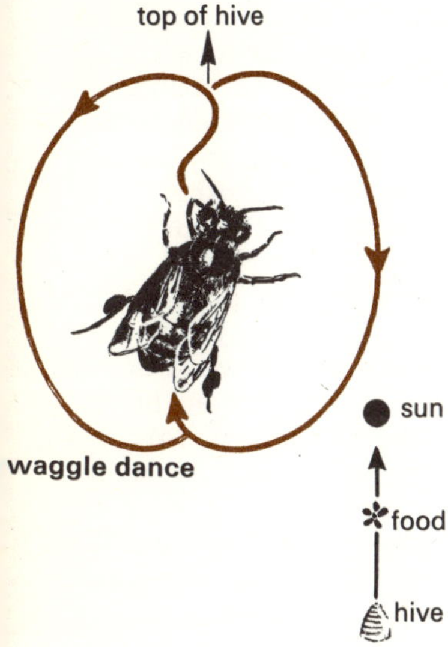

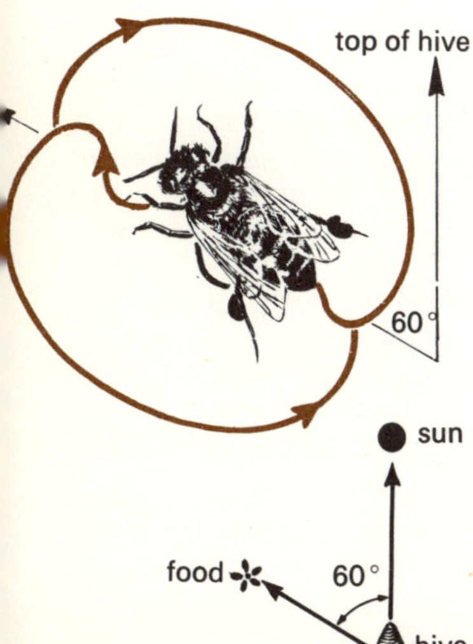

If the food is more than 90 m away, the dance becomes more complicated. If the food source is only a little more than 90 m away, the dance will be fast, but a bee which has just returned from a distant food source will dance more slowly. In addition, the pattern of the dance can tell other foragers the direction in which the food lies.

Both ants and bees show a form of social grouping which is typical of insect societies. There is instinctive division of labour. The society is held together by trophallaxis, and activities such as foraging and caring for the young are closely co-ordinated.

Vertebrate groups

Vertebrate groups are organized on a pattern different from that of insect societies. We have talked about two types of small social groups of vertebrates, the pair and the parent/young group, but many vertebrates spend a part of their lives in larger groups. In these groups, many of the features which are so typical of insect societies are completely absent. There is, apparently, no instinctive (innate) division of labour. Trophallaxis does not occur and, although there may be some food-sharing, it is more usual for each member of the group to find its own food and care for its own young.

If several chickens are kept together for a few weeks, one dominant member of the group will emerge. This chicken will be so dominant that it will be able to peck any other member of the group without being attacked. The organization of the group does not end there. There will also be a second bird which will be able to peck any bird apart from the dominant one. This 'peck-order' will extend throughout the group and at the bottom will be one bird who can be pecked by all the other group members, but which cannot peck back.

A similar system of social organization occurs in other vertebrate groups, including those of monkeys and wolves, although it may not be so clear-cut as the peck-order of chickens. Nevertheless, the dominance of one animal, usually a mature male, will be apparent. Dominant male baboons and wolves walk with their tails raised. Other animals approaching the dominant male usually do so with tails lowered—to raise them may provoke an attack. In this way, a dominant male uses a signal to show his rank. This type of social group within which the members are arranged according to their rank is called a *dominance hierarchy*.

Fighting Many examples of fighting behaviour in vertebrates have been observed. You can study some of these in the laboratory, but examples which you are unlikely to encounter include the fighting of rattlesnakes, oryx antelopes and wolves. These animals all possess weapons which could kill their opponents. The razor-sharp horns of an oryx are deadly, as are the teeth of a wolf, and the venom of a rattlesnake. But when these animals fight members of their own species, they do not use their deadly weapons to full effect. Male oryx stand side by side and exchange blows with the upper parts of their horns, as if they were fencing. They then pause, turn to face each other, interlock their horns and push head to head, until one of them gives ground. The fight then stops and the defeated animal withdraws.

Male rattlesnakes glide alongside each other. They raise their heads and push them together, each trying to push the other to the ground. The successful snake simply pins the head of its opponent to the ground for a few seconds and then allows it to glide away, defeated.

In some species, fighting ends abruptly when one of the animals submits (admits defeat). It may be unnecessary for the defeated animal to run away. Wolves show this type of fighting. A fight between two males ends when the weaker wolf turns his head aside, so exposing the side of his neck and throat. This could be fatal, for one bite here could cut through a large blood vessel. However, the victorious wolf does not carry through his attack, but allows his opponent to live on.

Aggressive behaviour plays a very important part in the establishment of a dominance hierarchy, but this is not the only reason for fighting behaviour. Fighting is also an expression of competition between members of the same species for food, shelter and nesting material. Animals may also fight for mates, to make sure that the animals which reproduce are the strongest individuals.

Whatever the reasons for fighting, each species has developed its own pattern of aggressive behaviour which seems to be inherited and not learned as the animal develops. Young animals of the same species raised in isolation will not only fight when brought together, but also their fighting will follow a pattern typical of their species. Experiments proving these points have been carried out with a variety of animals, including rats and various kinds of fishes.

It would seem that many vertebrates inherit patterns of aggressive behaviour which are typical of their

The seal on the left contracts his nose and submits to the bull on the right in a serious fight for dominance of the seal harem. Below: two young bulls in a practice fight.

species and which also avoid injury to individual animals. Injury or death of these individual animals would affect the chances of survival of the species as a whole.

Territory Aggression is an important part of territorial behaviour. At the beginning of the breeding season many male animals adopt a territory—a part of their surroundings which they will defend against members of the same species. Animals defend their territory in many ways. The main function of bird song may be to warn off other males of the same species. However, this is not always successful and sometimes the warning is ignored. If this happens, the resident bird attacks the intruder. If the intruder is a bird without territory it may stay to fight, but it normally flies away because a bird is always more confident on its own territory than it is on that of other birds.

Even the owners of neighbouring territories will fly away if attacked. It is in cases like this that you can sometimes see an interesting switch in behaviour. As the two birds cross the border between their territories their roles will change. The fleeing bird will turn and attack its pursuer. This rapid change of behaviour may happen again and again as they cross and recross the boundary between their territories. This *pendulum fighting* is a very common feature of territorial behaviour.

Sometimes a type of behaviour develops which seems to combine elements of both attack and escape. This compromise behaviour, which is called *threat*, usually occurs at the boundary of the animal's territory and is often enough to put off intruders. As a result, conflict is avoided. You can study threat behaviour in the stickleback. Refer to the investigation section for details.

Territorial behaviour is a means of ensuring survival of species in the following ways. First, it prevents overcrowding. Second, it ensures that there will be an adequate supply of food for the young. Some animals control quite large territories. Grazing animals, such as deer, may require 10 acres of forest to ensure their food supplies. Such animals occupy all suitable sites and those that arrive late in the area or are too immature or not aggressive enough to win territories cannot breed. In this way, the population is controlled and an adequate supply of food ensured. However, some cases are not so clear-cut. The territory guarded by gulls

A nesting gull threatens an intruder.

Below and far left: threat postures of a rat and the stickleback.

on a cliff-face may cover only a square metre, and be totally devoid of food. How, then, does this help to control populations and ensure adequate food? The territorial behaviour of the gulls limits the total number in the colony and thus, in turn, the number fishing in the sea nearby.

The social lives of vertebrates are not based entirely on competitive behaviour. There are advantages to be gained from just being together. Fish in shoals, birds in flocks, and mammals in herds all gain protection because there is safety in numbers of adults. A hawk is confused by the large number of birds in a flock and finds it difficult to select one individual as prey. Diving through a flock of birds at high speed is dangerous, too; a collision could injure the hawk. For these reasons, predatory birds tend to attack solitary prey, or those weaker individuals which straggle behind the main flock.

An animal sometimes does more in a group than it would do alone. This phenomenon is known as *social*

facilitation. For example, ants work harder and move more soil when they work with others than they do when they work alone. Many birds eat more in groups than they do when feeding on their own.

Co-operation In highly organized vertebrate societies we often see examples of co-operation between members. Co-operation arises when a problem cannot be solved by one animal alone. Many examples of co-operation between animals have been observed in the wild, and by casual observation of animals in captivity. You may have watched monkeys in a zoo picking parasites from each other's fur. Here is another example: a dolphin was tipped into a large pool for observation when, as it was being tipped, it hit its head on the edge of the pool and was knocked out. Dolphins are air-breathers and the animal would have drowned if the two dolphins already in the tank had not swum alongside the unconscious animal and lifted it to the surface. The stunned dophin was thus able to breath until it recovered.

In a laboratory experiment on co-operation, rats were kept separately in specially designed cages. Each cage had a floor which could be electrified, a platform at one end which a rat could press down to switch off an electric current, and a dish of food at the opposite end of the cage. First, each rat was placed in the cage with the current switched off and learned to feed from the dish. The current was then turned on and the rat learned that it could avoid an electric shock by pressing down the platform.

Pairs of trained rats were then placed together in similar cages. At first, they both fed together in spite of the electric shock. Furthermore, they both rested together, pressing down the platform to switch off the current. However, some pairs eventually learned to co-operate. One animal would depress the platform while the other fed. After a while, the waiting rat would bite or pull the tail of its feeding partner, which would return to the platform and allow the waiting rat to feed.

One of the most complex examples of co-operation between laboratory animals was shown by chimpanzees. Two chimpanzees were placed in separate cages. Each cage contained two coloured panels, perhaps red and green in one cage, blue and yellow in the other. All these panels had to be pressed in the right order before either chimpanzee was rewarded, which meant that the animals had to take turns in pressing panels, say, red

Co-operation betweer rats in a laboratory test.

followed by blue followed by green followed by yellow. The pair of chimpanzees solved this problem. In fact, if one animal was slow to respond, its partner would reach through the bars and shake or push it!

Animal groups on the move

The behaviour of groups often follows a routine. Members of groups may eat at the same time and sleep for similar lengths of time. Groups also tend to travel together.

Migration is a group movement to find new feeding or breeding grounds. Many animals migrate individually, relying on their own ability to navigate correctly. Other animals, notably locusts and swallows, migrate in groups.

Locusts fly actively but the direction of their migration is controlled by the wind which blows them along. Swallows migrate in a definite direction even if this means flying against the wind. However, the two animals have something in common: they both tend to remain in the group. If either of them becomes separated from the main migrating group, they will turn back towards it.

Animals derive a variety of advantages from migration. Swallows which migrate between Britain and southern Africa are able to breed in the temperate spring and summer climate of Britain, and to make sure of a constant supply of their insect food by returning to southern Africa during Britain's winter.

Some animals travel thousands of kilometres during migration. See what you can find out about the migra-

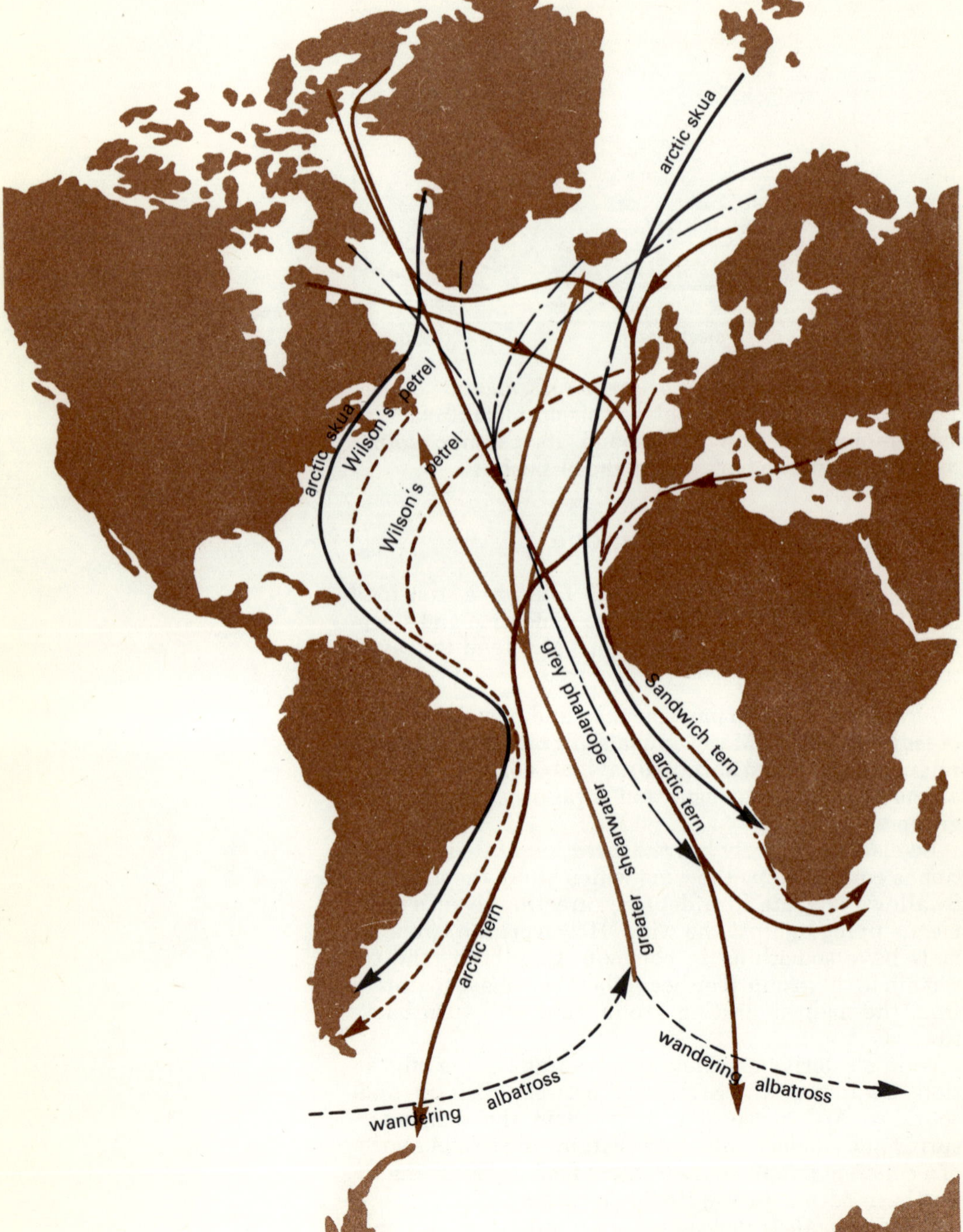
arctic skua
Wilson's petrel
arctic skua
Wilson's petrel
grey phalarope
Sandwich tern
arctic tern
greater shearwater
arctic tern
wandering albatross
wandering albatross

tion of some of the following animals: humming birds, butterflies, eels, salmon and redwing. What advantages do you think these animals derive from migration?

During migration, animals often navigate with incredible accuracy. Salmon spend several years of their lives at sea, but many return to breed in the river system in which they were hatched. Some birds can find their way home from places completely unfamiliar to them. One bird, a Manx shearwater, was captured on Skokholm, an island off the Pembrokeshire coast, and taken to Boston, in the United States, where it was released. It arrived back in Skokholm twelve and a half days later, having flown nearly 5000 kilometres!

How can animals find their way with such accuracy? The answer is: 'We don't know'. Many ideas have been put forward but we still have much to find out. It seems likely that animals use a variety of methods. These may include an inherited tendency to follow a particular route, guided by the sun or stars. Salmon may be sensitive to chemicals dissolved in the water of their own river system. But how a Manx shearwater can return to its nest-hole on the small island of Skokholm from a town in North America well outside its normal range is still a total mystery.

Human groups

Man has a more complex social life than any other animal. His high intelligence has led him to form highly organized societies. Man's intelligence shows itself in many ways and we will now look at three aspects of intelligence which are important in our social lives.

Communication Before any group can be organized the animals involved must be able to transmit and receive information. The ability to communicate is more highly developed in human beings than it is in any other animal. You will have already read about examples of communication in earlier parts of this book. These examples included courtship rituals, bird song, chemical communication, and so on. All the examples were of innate abilities to communicate.

Human communication is a combination of innate and learned activities, and it takes several forms.

A round trip of 20,000 miles is flown by the Arctic tern during its annual migration. Trunk routes for several long-distance fliers are shown in this migration map of the Atlantic.

What is this child expressing? How does his face change to suit his emotions?

Facial expressions can communicate a lot of information. Can you describe the expressions of fear, anger, or happiness? Can you think of any other examples of facial expression? The positions and movements of our bodies can also pass on information. Think what information would be passed on by someone carrying out the following activities: waving, pointing, clenching fists, wringing hands. Which of the activities are inherited and which are learned?

Language is the most advanced form of human communication. Some animals can communicate by sounds such as assembly calls, alarm signals, and aggressive song. Whales communicate by using a wide range of whistles and clicking sounds. As far as we know, human speech is the most complicated example of sound communication. Some other animals possess voice organs which are capable of producing a wide variety of noises. However, their brains are not well enough developed to enable them to learn and remember the noises in the way that we learn and remember words.

People use words in two very different ways. First, words are used to produce definite responses. For example, the words 'look out' usually cause the same response. In this case, the tone of voice is as important as the meaning of the words. Second, words may be used as symbols of objects, activities, or ideas. Words such as God, love, or law are symbols for very complicated ideas and their use indicates a high level of intelligence.

Information is passed on from one generation to the next by the use of speech. When man learned to speak he was able to pass on the benefits of his experience to his children. When the young grew up they, in turn, passed on this information to their children, who were able to add to it because of their own experiences. In this way, the amount of information passed on gradually increased. This process has been made much easier by the development of writing, which is itself a form of language. Written records mean that information can be stored or passed to people living a long way off.

Adaptability is a second important factor in the development of human society. In a primitive society a man must play many roles. He must be a hunter or farmer to provide food for his family. He must protect the family from enemies. If the climate is harsh, he must find or build a shelter. In more advanced societies there is a *division of labour* which means that the specialized tasks are divided among individuals. Some people will grow food, some will build homes, some will educate the children, and so on. Insect societies are very different but they also show a division of labour. In insect societies, the members are born with instincts which make them carry out their own particular jobs.

The Egyptians communicated information and ideas by hieroglyphs, or symbolic images. We use the alphabet.

Asparagus graders in an Australian canning factory. Can you imagine a more specialized job?

In a human society, a man may adapt himself to carry out one of a whole variety of jobs.

Division of labour helps an animal group to survive. In human society each person can specialize. He can learn a lot about a particular job and become very skilled at it. As a result, the necessary work of a group is carried out efficiently to the benefit of all. Would you rather be operated on by a full- or part-time surgeon?

Specialization also means that each job is carried out in the most suitable area. For example, food can be grown by individuals who live in areas with the best soil. Bricks can be made by those who live near clay. The farmer and the brickmaker can then supply each other with their products. In this way we have become very dependent on the behaviour of others, as we find whenever there is a strike.

Control of the environment When man was a primitive hunter he moved from one place to another in pursuit of animals for food and to find shelter from enemies and bad weather. As he gained control of his surroundings he was able to live in settled societies. He learned to grow seeds and keep animals. In settled societies he was able to build permanent shelters to protect him from the weather and from wild animals. His farming provided him with a constant supply of food. In these ways, increasing control of the environment led to the formation of societies and increased the chances of survival of individuals and the species.

Sub-groups of human society

You have learned that communication, division of labour, and control of the environment have all helped in the development of human societies. The complexity of our society is shown by the large number of smaller groups (sub-groups) of which it is made.

Some sub-groups are well-defined. For example, a football team is well-defined because we can easily recognize who belongs to it and who does not. Other examples of such sub-groups are schools, trade unions and regiments. There are other sub-groups which are more vague and harder to define. Examples include the working class, the middle-aged and the poor. Try to describe the members of each of these groups and you will soon see how difficult definitions are.

Another way of thinking about sub-groups is to consider their size. This may seem unimportant, but the size of a group decides how much an individual can be involved in its management. For example, it is easier, on the whole, to be elected to a village council than to a city council.

In the course of our lives we belong to many groups of different types and sizes. Try to think of the groups you belong to. Look for the reasons why you belong to each group. How much of your behaviour is determined by the groups you belong to?

One kind of human sub-group is common to practically all of us—the family. Some people believe that families are unnecessary. They imagine a society in which parents do not get married and in which children are brought up by the State. However, as things are at present the family remains the basic social group in human society. A growing child learns much of his behaviour from his family. The family is a very important influence when the child lives in a village and most of his relatives live nearby. The trend today is for families to become split up. Children grow up and leave the villages where they were born to find work and settle in the larger towns. Rehousing is another factor which breaks up families. The splitting up of a family means that the children will not now come under the influence of their parents only; they will come under the influence of people from other families. Families are also splitting up in another way: more and more parents are getting divorced. Do you think that the idea of the family belongs to the past?

This boy belongs to many sub-groups. How many can you spot in the picture?

Control of groups

You have already learnt that you belong to many groups of various types and sizes. In most of them your behaviour is controlled to some extent by the people who have authority over you. In a football team your actions are controlled by the captain. In a school your actions are controlled by the headmaster. Do you believe it is necessary for a group to have people in it who are in control over others in the group?

The control of groups is often arranged in layers. For example, in a school the headmaster controls the teachers; teachers control the pupils. In this way a school has a chain of control with the headmaster at the top and the pupils at the bottom. This organization of control is known as a *social hierarchy*.

What sort of person tends to gain control of a group? The type of person varies according to the group. In a street gang the leader will quite likely be the best fighter. There may be another member of the gang who has good ideas or a personality which people copy. But this person will still be under the authority of the gang leader.

In other groups, such as a factory, the position of a person in the social hierarchy will depend on many factors: intelligence, education, management ability, and the possession of special skills.

Hierarchy is not the only method of controlling groups. In some groups all the members are equal and decisions are made by asking the views of all the members in it. An individual who is not prepared to accept decisions agreed by most of the group may be excluded.

Another way of control is to apply sets of laws or rules. If a member of a group does not obey them, he may be punished or removed from the group. Removal may take the form of deportation to another country (exile), imprisonment, dishonourable discharge (from the army), etc.

To some extent your group behaviour is controlled by all the other members of the group. There are ways of behaving which are socially acceptable and we are encouraged to behave in these ways. Think of examples of behaviour which are either socially acceptable or unacceptable. Fashions also control behaviour. You probably know people who will do anything if it is fashionable. Taking drugs is fashionable these days and people who do not join in are often made to feel 'square' or 'yellow'.

Moral standards have a big effect on most people's actions. As we grow up we learn what is acceptable at the time. However, each generation has its own ideas about moral values. Not very long ago it was thought that pictures of naked people should not be shown in public. Today, plays, films and advertisements feature naked people.

Some young people are quite happy to take their parents' word for what is right and what is wrong. Others are determined to make up their own minds. People who disagree with the rules of society or its moral values will often get together to form their own groups with their own rules.

Think about your behaviour over the past 24 hours. How much of it was controlled by people with authority over you, by rules, by laws, by regulations, and by moral standards?

Mental illness and society

You read earlier that some people develop physical, mental, and emotional problems which make it difficult for them to fit easily into society. It has been accepted for some time that society should do everything it can to enable the physically handicapped to play an active part in the life of the community. We have hospitals for the sick, and training centres for the handicapped, although there is still plenty more we can do.

Society's attitude to the mentally ill is not so clear, however. The whole pattern of our society depends very heavily on co-operation between individuals. People with mental and emotional problems may find it difficult to co-operate with others and to obey the simple rules of society. The behaviour of such people may, in extreme cases, threaten the running of a community. In such cases we must take action.

What would you do? Basically, there are two points of view. One is that people who threaten society should be excluded—locked away, in fact—to protect the group. The other view is that such people are sick and need help. In some cases, of course, we have no choice. A dangerous psychopath must be excluded from society because the lives of other members of the community will be endangered if he is not. In other cases, it is far more difficult to decide on a proper course of action.

Society's attitudes about mentally and emotionally disturbed people are undergoing a gradual change, however. This change is long overdue and is happening too slowly, but it is definitely a change for the better. Until well into the 1800s, mentally ill people were caged like animals and exhibited for all to see. People would visit asylums then just as you may go to the zoo today. Even the Lunacy Act (1890) was designed, in the first place, to protect society. It did so by providing the legal power to separate mentally ill people from the rest of the community. How did people decide who should be declared insane and kept in asylums? There is little doubt that some spent their lives locked away who were not sufficiently ill to be treated in this way.

It was not until about 25 years ago that the public's attitude towards mental illness began to change. Mental illness is gradually being accepted as another form of sickness requiring specialist treatment. Mental patients are given much greater freedom today, being allowed to go home for weekends or away for holidays, going out of hospital to find work, and so on. Individuals may now enter hospitals as voluntary patients if they feel that they are under serious mental stress.

There is still much to be done, however. Many mental hospitals are unsatisfactory. Some of the buildings are bleak and the accommodation unpleasant, although large sums of money are now being spent to correct these deficiencies. At the same time, the treatment of mental illness has changed. It is no longer enough to keep patients quiet and securely enclosed. For one thing, the problem itself is too big to lock away. There are more than 200 000 beds in hospitals throughout the country occupied by mentally ill and mentally deficient people. So the aim now is to treat as many patients as possible so that they may return and take their place in society.

Some mental hospitals are desperately overcrowded. How would you like to live like this?

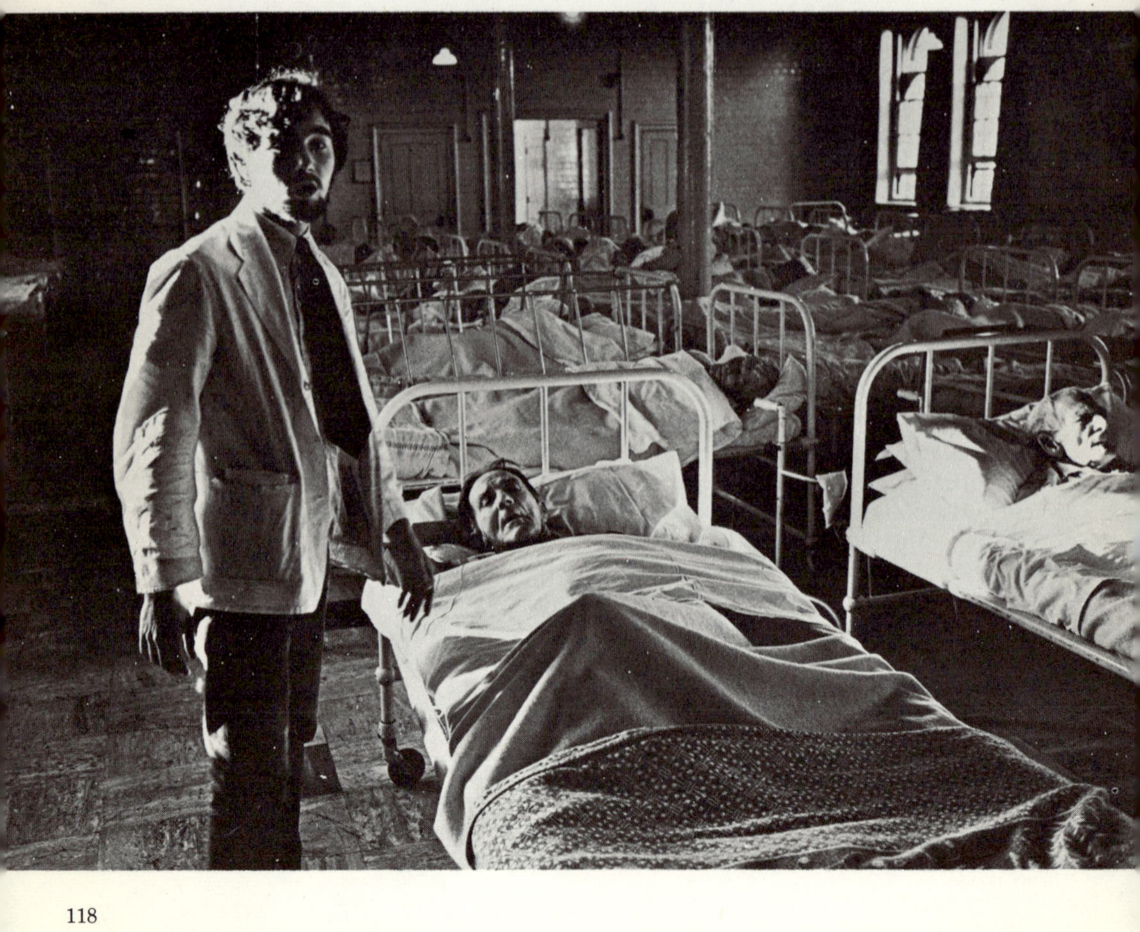

Investigations

In the section which follows you will find experiments you can carry out to investigate further many of the topics discussed in the earlier part of the book. You will also be able to build on these to develop your own experiments, and much of the simple apparatus is adaptable for this. Some of the experiments can be carried out in pairs, others are better as team projects, just as some will show immediate results and others are for observation over a period of time. In any kind of investigation, however, it is important to remember to record your results, at all stages, clearly and accurately. Discussion topics are also suggested and these may give rise to further areas for debate and research for which helpful addresses and references are given.

Apparatus and materials

Laboratory apparatus

Petri dishes
beakers
clamps and stands
test-tubes
measuring cylinders
Bunsen burners
Dreschel bottle
stop-watch
pie dishes
stirring rods
drinking bottle
chemical balance
specimen tubes
scalpel or knife
forceps
callipers
audio-frequency signal generator
cathode ray oscilloscope
24 hour motor
bench lamp or torch
electric bell
battery for bell
aquaria
mirror
glass tubing
plastic tubing
rubber bungs
corks
plastic cage

Chemicals

distilled water
acetic acid
quinine
sodium chloride
chloroform
formic acid
calcium chloride
benzene
ethyl acetate

Living specimens

planaria
Tubifex
woodlice
ants
Dytiscus adult and nymph
earthworms
brine shrimps
stick insects
caterpillars
earwigs
mosquito larvae and pupae
aquatic plants
sticklebacks
gerbils
fly maggots
cockroaches
beetles

Other apparatus and materials

filter paper
Plasticine
sugar
sea water
meat extract
yeast
cotton wool
perforated zinc
enamel or wooden containers
glass sheets
flower pot
strips of wood
16–18 swg wire
glue
pencils
hamster wheel
ice
hot water bottle
blotting paper
cardboard boxes
coloured paints
elastic bands
polystyrene tiles
nylon netting
pins
cotton
brass tubing
engine oil
foam rubber
kymograph or smooth paper
paper (black and white)
newspaper
cylindrical tin
mild steel rod
metal sheet
electrical wire
nail
insulating tape
sand or gravel
opaque glass
model of male stickleback
hay and peat
clove oil
cellulose paint or nail varnish
leak-proof tray
liver or other meat
soil
strong-smelling substances

1 Some aspects of human vision

The pupil is a hole which appears as the dark circle in the centre of the iris which is the coloured circle in the eye. Look at someone's eyes. Note that normally both pupils are the same size.

Ask your companion to cover one eye with his hand and then to look at a light for a few minutes. At the end of this time ask him to remove his hand and as he does so look at the pupil and iris of the eye which was covered.

1 How does the size of the pupil alter when it is exposed to the light?
2 Light must pass through the pupil in order to reach the retina. Why is the size of the pupil different in light and dark conditions?
3 Why should you never look directly at the sun?
4 What artificial aids are commonly used to control the amount of light entering the eyes?

There are no light-sensitive cells where the optic nerve leaves the eye and hence this region is known as the blind spot. The existence of this region can easily be demonstrated.

Hold the book about 40 cm from your eyes. Close your left eye and look at the cross with your right eye. Now move the book slowly towards you and notice that at a certain point the spot disappears. In this position the image of the spot is falling on the blind spot. Can you work out the approximate position where the optic nerve leaves the eye?

Although we do not realise it, only a very small part of what we see at any time is clearly in focus. The eye constantly focusses on different objects. To illustrate this, look out of a window while holding a pencil between your eyes and the window. Focus on some distant object outside and notice how the pencil is out of focus. Now focus on the pencil. Are the distant objects now in focus?

The process by which the focus of the eye changes is called accommodation and it is accomplished by muscles which change the shape of the lens.

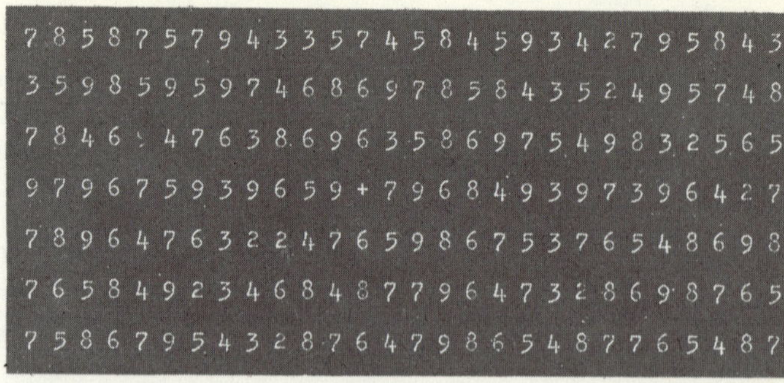

You will need

pencils
paper
Plasticine

There is a small area of the retina which is especially sensitive to light and where only cones are found. This area is called the fovea centralis or yellow spot. When we look at an object, the eyeball muscles move the eye until the image of the object falls on the yellow spot. Of course, other objects can be seen at the same time —but not so clearly.

You can demonstrate this fact by copying the figure above onto a sheet of paper. (You need not use the same numbers.) Ask a friend to cover his left eye and concentrate on the cross while the paper is about 10–15 cm from his right eye. Point to each number in turn, starting from the cross and working outwards. Cross off each number as it is correctly identified and continue until your friend cannot identify any more numbers without moving his gaze from the cross. Draw a line between the identified and unidentified figures. This experiment shows your friend's field of clear vision when the image of the cross is falling on the yellow spot of the right eye. Prepare another sheet of paper and with the help of your friend, find your own field of clear vision and compare your two results.

Use a piece of Plasticine to hold a sharp pencil upright on a table. Look along the table at the pencil so that it is about 30 cm from your head. First look at the pencil with one eye closed and then with both eyes open. When did the pencil appear more solid?

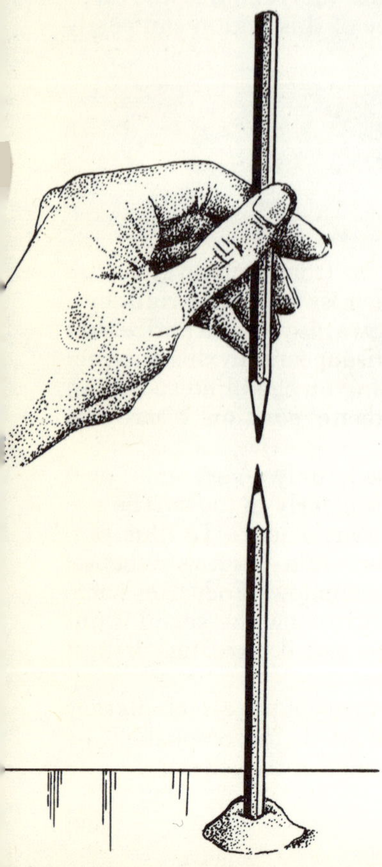

Now close your left eye, hold another pencil in your left hand, bring the pencil down slowly and attempt to make its point touch the point of the pencil on the table. Repeat this twenty times and record the number of hits you score. Now repeat the experiment with both eyes open and with the right eye closed. When you look at the pencil with only one eye, your vision is monocular as opposed to binocular vision when both eyes are used. Can you see any advantages of binocular vision?

2 The limitations of the senses

You have read in the text how our senses are used to gain information about the environment. The accuracy of this information is important because much of our behaviour is based on it. Here are a few simple experiments which show some of the limitations of the senses.

1 When we see something, the sensation of the image is formed in the brain and not in the eye. What you see, or at least what you recognize, depends largely on your previous experience. The brain is not infallible and in some cases may be easily misled. A few of the commonest examples of optical illusions are illustrated here.

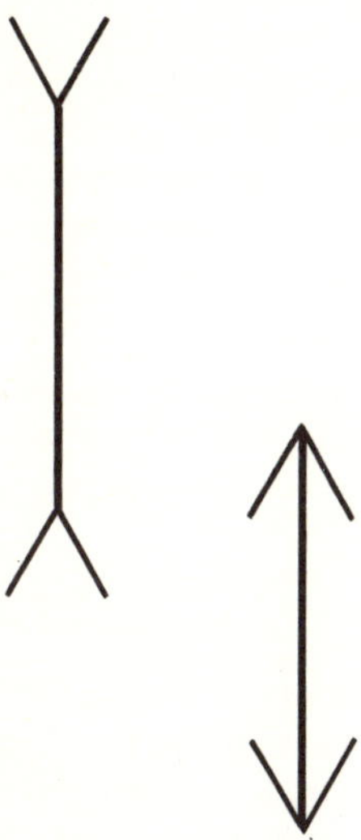

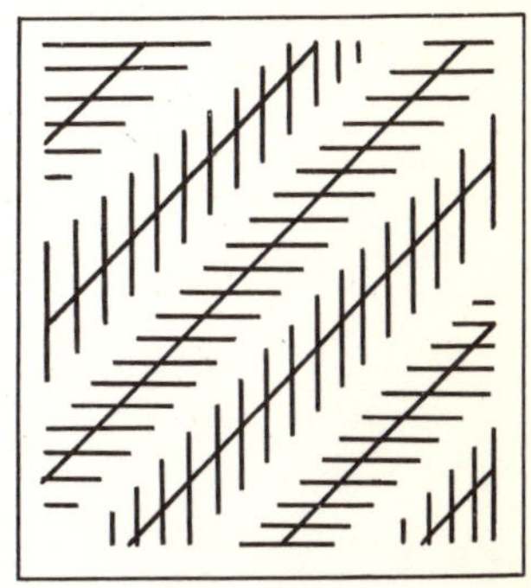

It is hard to believe that the two vertical lines above are the same length, or that the diagonals on the figure on the right are really parallel lines. If you look at the circle, and move the book slightly, it is hard to believe it is not rotating.

2 How sensitive is our sense of taste? What is the weakest solution of sugar which we can detect? Prepare a solution of 10 g sugar in 100 cm^3 water. Now prepare a series of weaker solutions as follows. To half of the solution add 50 cm^3 water to make a half strength solution. To half of the latter solution add 50 cm^3 water to make a quarter strength solution. Repeat this process to give solutions of ten various strengths. Make sure that all solutions are carefully labelled.

Allow the subject to taste the weakest solution and then some water. Can he identify the taste of the sugar? Use the other solutions in the same way so that you are increasing the strength each time. What is the weakest solution that the subject can identify? You can improve the experiment by sometimes giving water instead of sugar solution. The subject can be told that you are going to do this but on any given test he will not know if he is receiving sugar or not.

Try this experiment using salt instead of sugar.

3 Another threshold of sensitivity experiment can be carried out on the skin. You will need a subject and a pair of callipers. Ask the subject to close his eyes and then touch his palm with one or two points of the callipers. When using both points, make sure that they both touch the skin at the same time. The subject should say whether he is being touched by one or two points. If he can successfully distinguish these two events, put the points of the callipers closer together and repeat the procedure. Continue in this way until the subject cannot distinguish between one and two points and measure the distance between the points of the callipers. Repeat the experiment on various parts of the body, for instance, the fingertips, lips, cheeks and forehead to find out which is the most sensitive part of the body in this respect.

4 The skin is also sensitive to temperature, but how reliable is this sense? Take three beakers and fill one with hot water, one with cold and one with a mixture of the two. Place a finger of one hand in the hot water and a finger of the other hand in the cold water and leave them there for a minute. Now put both fingers in the warm water. What happens? Can you explain why?

5 Of all the sounds which can be produced, we can hear only those within a certain frequency range. What is this range? To find this out, you will need the help of a physicist and you will need to borrow an audio-frequency signal generator and a cathode ray oscilloscope from a physics laboratory. The signal generator produces sound of various pitches or frequencies. The higher the pitch of the note, the higher the frequency. From the scale on the generator, you can read off the frequency of any particular sound produced. The physicist will use the equipment to produce sound of increasing and decreasing frequencies. Record the points at which the frequencies are too high or too low for you to hear. The oscilloscope will show that sounds are still being produced although you cannot hear them. Plot a graph of the highest and lowest frequencies audible to each member of your group.

You will need

- beakers
- sugar
- salt
- stirring rods
- callipers
- audio-frequency signal generator
- cathode ray oscilloscope

3 Behavioural responses to stimuli

The brine shrimp

Obtain from a tropical fish shop some eggs of the brine shrimp (*Artemia salina*). Hatch out the eggs by following the instructions supplied with them. Place 1 or 2 shrimps in water in a small glass tube. Stopper the tube and hold it horizontally. Make sure that the direction of the light falling on the tube is from above and observe and record the position in which the shrimps swim.

In a darkened room, shine a torch on the tube, first from the side and then from underneath. Observe and note any changes in the swimming position of the shrimps.

What do these results tell you about the way the direction of light affects the swimming position (orientation) of the shrimps?

You will need

glass tube
stopper to fit glass tube
torch (or similar light)
aquarium (to hatch shrimp eggs)
sea-water
yeast (to feed shrimps)

Planaria

Planaria are flatworms and can be obtained from specialist dealers or caught in freshwater ponds or streams.

Obtain a piece of clear plastic tubing about 5 cm long and about 5 mm in diameter and cut it lengthwise so as to form two small troughs.

Put a trough on a piece of white paper and make sure that it is evenly illuminated by overhead lighting. Put a few drops of water in the centre of the trough and place a flatworm in the water. Observe and record the direction(s) of its movements.

Soak a small piece of filter paper in strong sodium chloride (salt) solution and using forceps, slide the paper along the trough until it just touches the edge of the water. In this position the salt will slowly diffuse into the water. Observe and record any changes in the direction of the flatworm's movements. The results of experiments with one animal may not be reliable so

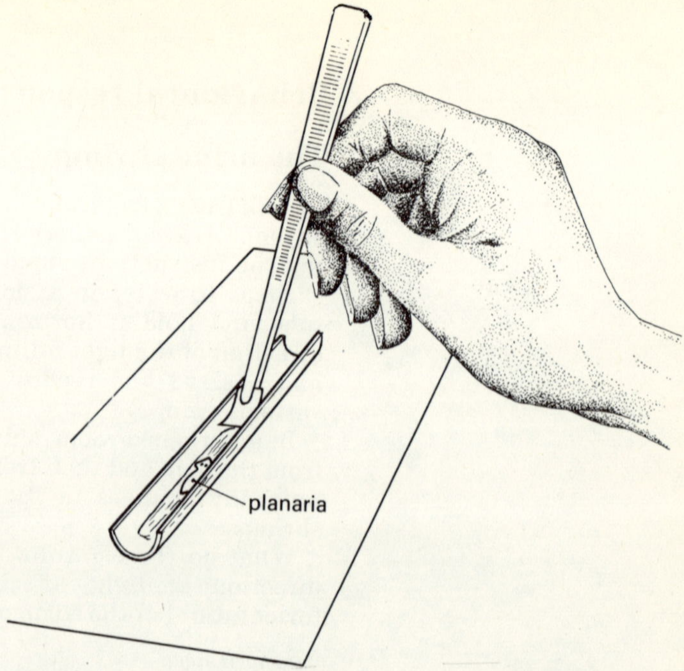

clean the trough with distilled water and repeat the experiment a few times using different planaria.

The other trough may be used in a similar way to test the response of planaria to another chemical such as acetic acid (vinegar), quinine, or chloroform.

1 Are planaria sensitive to any of the chemicals tested?

2 How do they respond to the chemicals?

3 Can you think of any reasons why some chemicals might be avoided?

4 As you performed this experiment, did it occur to you that flatworms might respond to filter paper soaked in distilled water? It should be easy for you to check this possibility.

You will need

plastic tubing (5 mm diameter)
white paper
distilled water
filter paper
forceps
sodium chloride solution
acetic acid
quinine
chloroform
aquarium (for planaria)
live *Tubifex* or fresh liver (to feed planaria)

The woodlouse

Collect about twenty specimens of woodlice of the same species. A common British woodlouse is *Oniscus assellus* which has a grey body with yellow markings and which does not roll up when touched. These may be found under wood, stones and other debris.

For this experiment a choice chamber can be constructed in the following manner. Obtain a container, such as a wooden box or enamel dish, with dimensions about 25 × 15 × 5 cm. Divide the container into two equal sections by drawing a line across its width at the centre. At one end of the container, place a Petri dish containing wet cotton wool; this is to provide a humid atmosphere. At the other end, place a Petri dish containing anhydrous calcium chloride which will absorb water vapour from the air and so produce a drier atmosphere.

Cut out a piece of perforated zinc with the same width and length as the interior of the container. Using pieces of Plasticine as supports, place the zinc over the dishes so that there is a small gap between the zinc and the dishes. Use masking tape to cover any gaps between the edges of the zinc and the walls of the container.

Put the woodlice on the zinc so that they are in a row along the line which divides the container into

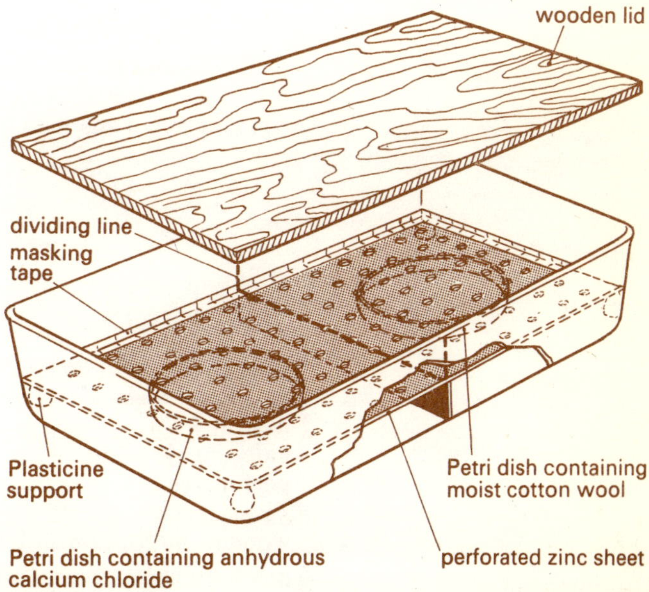

two. If this proves difficult because the woodlice move around too much, you can slow down their movements by putting them in a refrigerator for a short time.

Put a wooden lid over the container and do not remove the lid for one hour. After this time, remove the lid and using the central line as your criterion, count the number of woodlice in the dry half and the number in the humid half of the container. If there are other people repeating this experiment, you should combine your results, but if not then you should repeat the experiment a few times.

The experiment should now be repeated but using a glass cover instead of the wooden lid. Make sure that the interior of the container is evenly illuminated.

1 When woodlice are in the dark, do they prefer a humid or a dry environment?

2 Is this preference the same when the woodlice are in the light?

3 Repeat experiment using maggots instead of woodlice. How do they respond to illumination—do they prefer the light or the dark?

4 In which of the following environments did you find your woodlice?

a light and dry
b light and humid (damp)
c dark and dry
d dark and humid.

5 Is humidity or illumination more important to woodlice? Could you adapt the apparatus in the present experiment to answer this question?

You will need

container about 25 ×15 ×5 cm
two Petri dishes
cotton wool
calcium chloride
perforated zinc
Plasticine
wooden lid for container
glass sheet to cover container

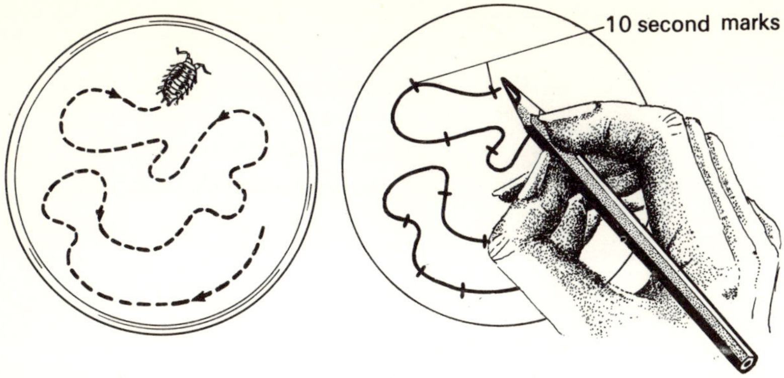

In the previous experiments you have studied the behaviour of woodlice in conditions where they have a choice of environment. In this investigation, you will be studying the reaction of a woodlouse in a constant environment and comparing this with its behaviour in a different environment.

Set up two Petri dishes with filter paper in the base. In one the filter paper should be moist, in the other it should be dry.

On a piece of plain paper, draw a circle the same size as the Petri dishes. Place a woodlouse in the moist petri dish and replace the lid. Observe the behaviour of the woodlouse and, using a pencil, copy the pattern of its movements into the circle you have drawn. Continue the recording for three minutes, marking off the line which you draw at ten second intervals. Work out the average speed of movement of the woodlouse. For increased accuracy, the experiment should be repeated several times, using different woodlice.

Repeat the above procedure using the dry Petri dish and compare the results. What conclusions can you draw from this investigation?

Can you devise a similar experiment to compare other contrasting environmental conditions and their effect on behaviour? For example, what about standing one Petri dish on a warm hot water bottle and the other on a layer of ice?

You will need

Petri dishes
filter paper
paper
pencil
stop clock
ice
hot water bottle
(or suitable container)

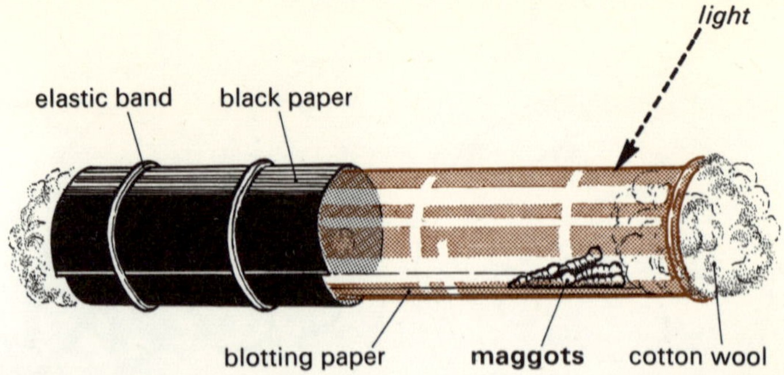

The fly maggot

Maggots can be bought at a shop which sells fishing bait.

Obtain a piece of glass tubing about 20 cm long and 3 cm in diameter. Cut a strip of blotting paper as long and wide as the tubing. Moisten the paper and lay it in the tubing (as shown above) so that it sticks to the glass. Plug each end of the tubing with cotton wool. One end of the tubing must now be darkened and this may be done by inserting half the tubing through a hole cut in a cardboard box. Alternately you can cover one half of the tubing with black paper held in place by an elastic band. Shine a torch or similar light on the end of the protruding tubing.

Remove the wool from the 'light' end of the tubing and put about ten fly maggots (larvae) just inside and replace the wool. Observe the movements of the maggots and every three minutes count and record the number of maggots in the 'light' end of the tubing. It is important to record your results for this and other experiments in a neat and clear manner.

1 Are fly maggots sensitive to light?

2 How do fly maggots respond to light?

3 Fly maggots live in rotting plant and animal matter. How might their response to light help them to survive?

4 Can you find out if adult flies respond to light in the same way as their larvae do?

You will need

glass tubing about 20 cm $\times$3 cm
blotting paper
cotton wool
cardboard box (or black paper and elastic band)
torch (or similar light)

4 Food-getting behaviour

One of the basic needs controlling animal behaviour is the need for food, and for many animals, food-seeking and food-getting form a major part of their activity.

1 Aphids are a suitable subject for investigation. Obtain six pie dishes and, using good gloss paint or enamel, paint them red, blue, yellow, green, black and white. Place a shallow layer of water in the bottom of each dish and place the dishes in a row close to a line of beans or in any area where aphids abound. After 24 hours, collect the dishes and count the number of aphids trapped in each. Repeat on several days. What conclusions can you draw from these results, if any? Can you think of a useful extension to this experiment?

You will need

six pie dishes
six coloured paints

2 During the summer months, it is often possible to obtain adult and nymphal forms of the water beetle *Dytiscus marginalis*, either from local ponds or biological suppliers. Place adult and nymph in separate small aquaria. Attach a small piece of liver or other meat to a piece of thread and draw it slowly through the water in the aquarium. Observe responses. One day later, repeat the experiment, but substitute a cube of Plasticine of similar size and colour for the meat. Observe reaction. Add a little solution of meat extract to the water of the aquarium, and repeat the procedure with the Plasticine. What are the responses of *Dytiscus* to the presence of food? Are the responses of the adult and nymph the same? Do the responses occur in any particular order?

You will need

small aquaria (two)
liver
cotton
plasticine
meat extract (e.g. Bovril)

3 Obtain some planaria. Place one planarian in a pool of water in a Petri dish and, using the technique described for use with the woodlice, trace the patterns of movement of the planarian, marking off the line at one minute intervals. Add one drop of Bovril solution to the water. Repeat your recording. Has the animal changed its rate of movement, or of turning?

Now place several planaria in a clean Petri dish in shallow water. Allow one hour for the animals to settle. Place a strip of paper soaked in liver extract or Bovril into one side of the dish. Is there any detectable change in the behaviour of the animals? Repeat experiment using a small piece of liver.

You will need

Petri dish
paper
pencil
stop clock
meat extract

5 Defensive behaviour

The stick insect, *Dixippus morosus*, has no weapons such as teeth, claws or sting organs with which it can defend itself. However it does have ways of protecting itself from its enemies.

If you look for a stick insect which is amongst a bunch of twigs, you will realise that it is very well camouflaged. The shape and colour of its body closely resemble those of a twig (left), and it has ways of behaving which also help to protect it.

The adults are able to change their colour. You can demonstrate this fact for yourself if you use two equally dark adult specimens. Put one in the dark and put the other on a dark background but one which is well illuminated. Leave them for a few hours and then compare their colours. What difference can be seen? What value does the colour change have for the insect?

If you keep some stick insects and observe them regularly, you will sometimes find one which appears to be dead. For instance, one may be lying on the floor with its first pair of legs stretched forward and the others stretched backwards so that it looks like a single twig. The insect appears to be 'feigning death'. Can you think how this behaviour will help to protect it from its predators?

One obvious way in which an animal can protect itself is by hiding. If you examine an old wall you will find many small animals living in its cracks. Turn over a stone or piece of wood and usually you will see a frantic rush for shelter.

1 The sense of touch plays a part in the hiding behaviour of some insects, such as the earwig. When an earwig is moving about, its body surface may come into contact with the sides of a crack. The earwig responds to this contact by slowing down or even ceasing its movements.

It could be argued that the earwig behaves in this way because it does not like the light and prefers the darkness of the crack. You can check this for yourself in the following way.

Put an earwig on a piece of white paper and cover it with a beaker. Shine a light on the beaker and observe the behaviour of the earwig, which now cannot escape from the light. Where does the earwig come to rest?

You will need

white paper
beaker
bench lamp

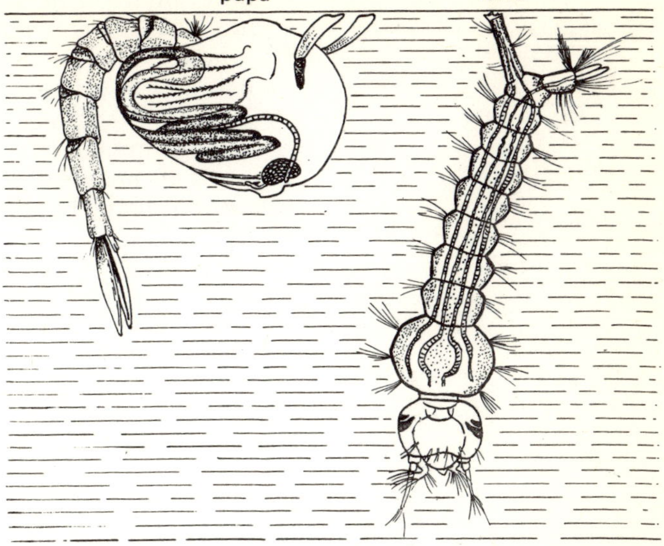

Immature stages of the mosquito.

2 When someone approaches you from behind, it may well be their shadow which tells you of their presence. A shadow may be that of a predator and so some animals show a response to shadows. This response can be clearly seen in the larvae and pupae of mosquitoes.

Collect some of these from stagnant water where they are found hanging from the surface film with their heads downwards. Why are they found in this place and in this position?

Put the larvae and pupae in a large beaker with a strong overhead light shining on it. Give the animals time to settle down and then move your hand between the light and the beaker so that its shadow passes over the surface of the water. How do the larvae and pupae respond to the shadow? Repeat the shadow-making action several times. Do the insects continue to make the response? If the response ceases, can you think of a reason for this? You may also like to find out if the response is shown:

a when the hand is moved very slowly over the water.

b when there is no strong light over the water.

Try to account for any findings.

You will need
beaker
bench lamp

3 Caterpillars of both butterflies and moths show a variety of responses when they feel themselves to be threatened.

Obtain a few caterpillars of as many different species of butterfly and moth as possible. If you live in a suitable area, you can collect these, but in the interests of conservation, it is more desirable that your specimens should be obtained from one of the specialist breeders. Suitable species include Geometrid moths, the Puss Moth, hawkmoths, tropical silkmoths and many others. The dealer may be able to advise on suitable species from the stocks which he has available.

Place individual caterpillars on twigs and observe carefully their responses to a variety of threats. These can include tapping the twig, switching on a bright light, passing a shadow over them, and in any other way you can contrive which avoids actual injury (gently touching the caterpillar with a paintbrush would do no harm). You may prefer to record some of the responses by photographs or sketches. Can you see any value in these responses? For example, is there any correlation between the pattern of colouring of the caterpillar and the type of response produced?

The caterpillars used in this experiment should not be discarded. Treat them carefully, keep them supplied with food, and attempt to rear them through their whole life cycle. If you are successful, you may even be able to observe the adults pairing and egg-laying.

4 Bees, like other social insects, will react aggressively if they feel that their colony is threatened. If there is a bee-hive anywhere in the vicinity of your school, you can investigate this aggression. It is desirable that the experimenters should wear suitable protective clothing.

The basic procedure in this experiment is to threaten the entrance to the hive with a series of small balls of cotton wool which have been treated in various ways.

Make 10–15 balls of cotton wool of 2–3 cm diameter and wrap them in muslin. Leave some in their natural state, colour others, place some in glass dishes with different small mammals for a few hours to pick up the animals' odour. Place drops of strong-smelling chemicals on others. Hold these cotton wool balls, in turn, on long wires, and place them a few centimetres from the entrance to the hive. In the first series of tests, the balls should be held motionless, while in a second series, they should be shaken.

Record your results. If any bees sting the balls, the stings will be left behind and may be counted. What conclusions can be drawn from these investigations? Do bees respond more aggressively to moving objects than they do to motionless objects? Do some colours or odours produce more aggressive responses?

You will need

cotton wool
paints
strong smelling chemicals
wire
muslin

6 Rhythmic behaviour

An actograph is a piece of apparatus which is used to investigate the activity of an animal during 24 hours. You can construct a simple actograph as follows.

Obtain about 50 cm of 16–18 swg wire. Take an old chemical balance from its case and discard the scale pans. Solder or glue the wire to the top of the beam so that about 30 cm of wire project from one end of the beam. Make a box about 8 × 4 × 4 cm using polystyrene ceiling tiles. Cover the open top of the box with nylon net attached with pins. Suspend the box from the wire.

Next you must construct a recording drum like the one shown in the diagram. Make the sleeve (D) from a 5 cm length of brass tubing which will fit tightly over the spindle in the 24 hour motor (E). Drill and tap the sleeve 1 cm from one end to accept a locking screw (F) and solder the other end to the motor spindle. Obtain a cylindrical tin about 12–15 cm in diameter and a mild steel rod, with a diameter to fit the sleeve and about 5 cm longer than the height of the tin. With a drill the same diameter as the rod, make holes centrally in the

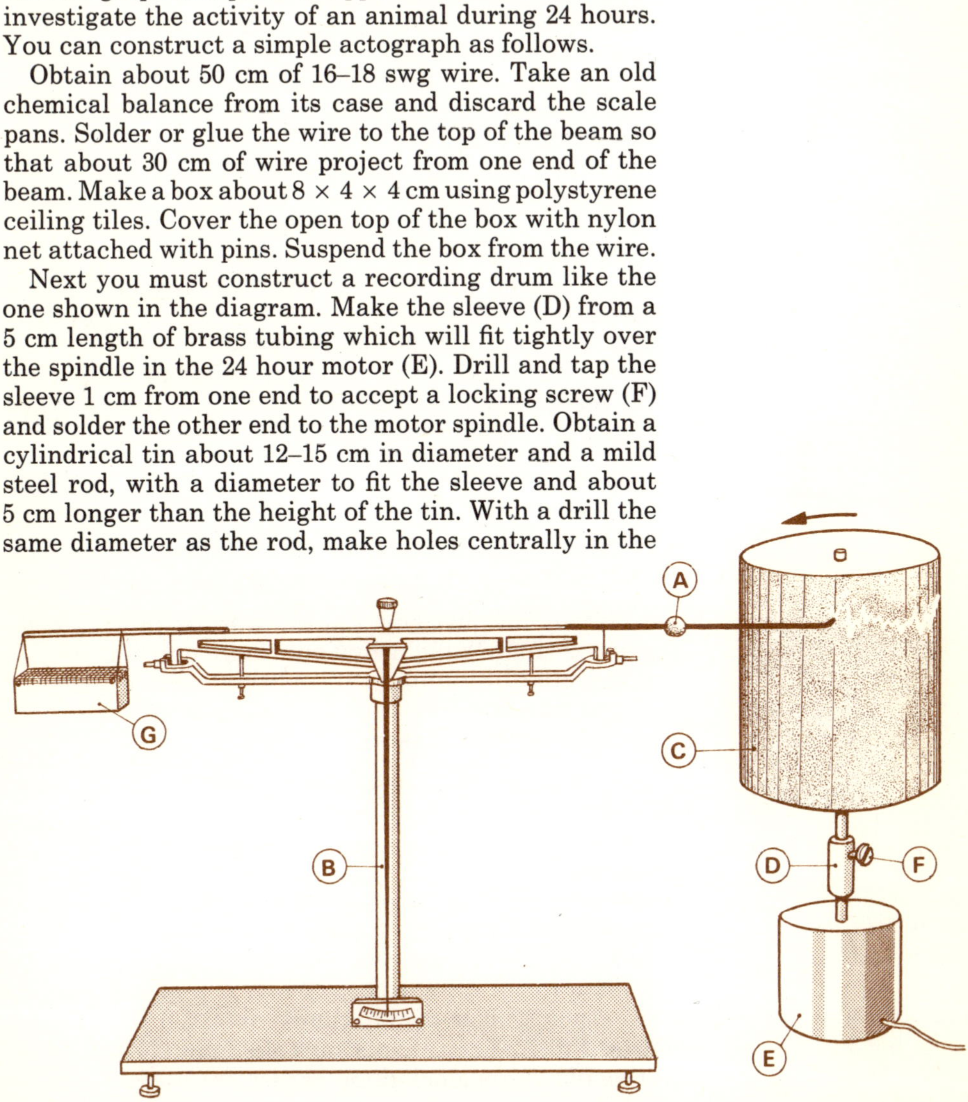

base and lid of the tin. Insert the rod through the holes and solder it to the tin so that 5 cm of the rod is projecting from the base. Insert this length into the sleeve and secure it with the locking screw.

Obtain some kymograph, or other smooth paper, which must be cut so that it fits around the drum and overlaps by at least 1 cm. Glue these overlapping edges so that the paper fits tightly to the drum. Bubble gas through a Dreschel bottle filled with benzene which connects with a Bunsen burner. This produces a smoky flame which is used to deposit soot on the smooth paper. To do this, rotate the drum at a moderate speed, in a horizontal position a few centimetres from the flame until it is thinly covered with soot.

Find the weight of the insect to be used and put that weight in the centre of the box. Put Plasticine at A on the wire so that the balance pointer (B) comes to rest on the zero mark. Replace the weight (in the box) with the insect and position the recording drum so that the end of the wire just touches the smoked paper. As the insect moves around the box, the balance of the apparatus will be disturbed, and the wire will move up and down against the smoked paper on the revolving drum. The result of this will be a record showing when the insect is active. The time of activity can be established by marking the paper into 24 (hours).

1 Use a stick insect and find out if its activity is mostly nocturnal or diurnal.

2 Repeat this experiment with other small animals such as bees, cock-roaches and beetles.

3 If you find that any of these animals shows a clear pattern of rhythmic behaviour (more active by day or more active by night) see if you can disturb the rhythm by varying conditions as follows. Repeat the experiment for several days with the apparatus:

a illuminated at all times
b in darkness at all times

You will need

50 cm of 16–18 swg wire
chemical balance
polystyrene tiles
nylon netting
pins cotton glue
Dreschel bottle
benzene Plasticine
5 cm brass tubing and a locking screw
cylindrical tin 12–15 cm diameter
mild steel rod Bunsen burner
kymograph or other smooth paper
twenty-four hour motor (can be obtained from Proops Bros. Ltd, 52 Tottenham Court Rd, London WI)
or an old alarm clock fitted with 2:1 reduction gearing

7 Variation in human reaction times

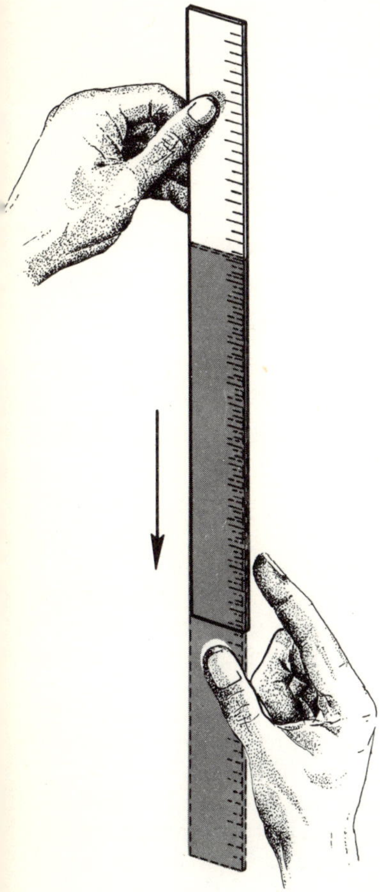

When we respond to a stimulus, there is always an amount of time (reaction time) between the reception of the stimulus and the performance of the response. Why is this so?

In the following experiment, you can find your reaction time in a simple stimulus—response situation. You will be the subject and you will need a companion to be the operator. The operator holds a ruler between the subject's fingers as shown in the diagram. Without warning, the operator drops the ruler, which the subject has to catch as quickly as possible. Record the distance the ruler drops before it is caught and calculate the reaction time by using the following formula.

$$\text{Reaction time (secs)} = \sqrt{\frac{2 \times \text{distance (cm)}}{960}}$$

If you repeat the operation ten times, you can find out if your performance improves with practice. Calculate the average reaction time by inserting the average distance (which the ruler falls) into the formula given.

Now repeat the experiment with other subjects and compare their performances. Is there much variation in reaction times?

Reaction times are very important when we consider road safety. For instance an alert driver travelling at a speed of 32 m/s (70 mph) will need an overall stopping distance of 96 m. Of this distance, 21 m are required for thinking and 75 m for braking.

You will need

ruler

8 Learning and memory

In this book you have read that one type of learned response is known as the *conditioned response*. This is a response which is brought about by a stimulus other than that which normally produces it.

You can produce a conditioned response in fish in the following way. Put some fish in an aquarium and when you feed them make sure that you always put the food in the same place on the surface. The fish will soon begin to gather quickly at this spot when they are fed.

Once this behaviour is established, begin the conditioning as follows. Each time you put food on the water, accompany your action by gently tapping the aquarium. Try to eliminate all other actions (such as lifting the lid of an aquarium) which might occur before feeding. After repeating this procedure every day for about three weeks, tap the aquarium but do not present any food. If the fish swim to their feeding position, they will be showing a conditioned response. In other words, the fish will have learnt to respond to the tapping. If your fish do not show this response, then continue the conditioning procedure until they do.

1 Do you think that it would be unwise to tap the aquarium at other times than feeding, and if so, why?

2 Why is it necessary not to accompany feeding with some other action like lifting an aquarium lid?

Try to think of examples of conditioned responses in your own behaviour.

Mirror drawing

This experiment is to find out if the learning of a skill when using the right hand is of any use in the future performance of the skill when using the left hand.

The skill involved is 'mirror' drawing and you will need to construct the electrical circuit shown in the diagram. The star shape on the metal sheet is made of plastic insulating tape. Stick the tape down and then, using a sharp knife, cut it so that it is 5 mm wide.

Arrange the tin lid, a mirror and a screen as shown. When the apparatus is used, a subject sits in front of it and observes the star in the mirror and makes sure that the screen prevents a direct view of the star. The subject takes the stylus and attempts to trace the star-shaped pathway in a clockwise direction. Each time the stylus leaves the plastic and touches the metal, the bell will ring and this is recorded as an error. The time taken to complete the circuit is also recorded.

The procedure for the experiment is as follows.

1 Divide the number of subjects available into two groups to be called the experimental group and the control group. Try to ensure that the groups are as similar as possible. Divide the subjects into pairs so that each pair contains an experimental subject and a control subject.

2 Each pair uses the apparatus thus:

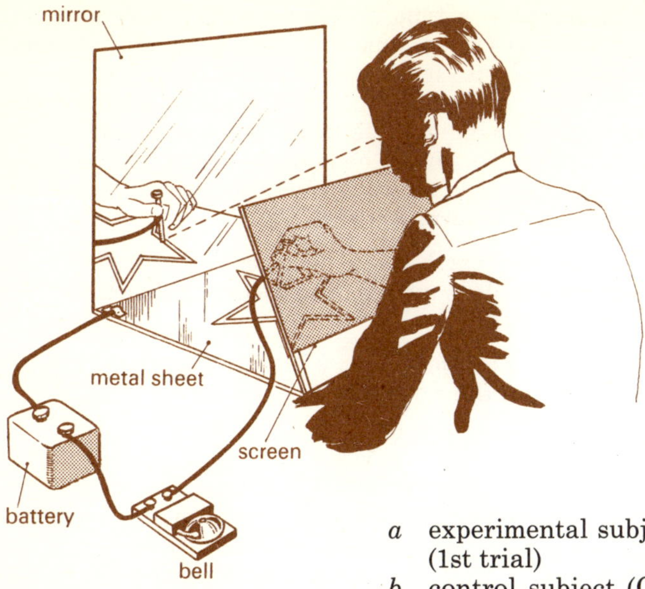

You will need

metal sheet 20 × 20 cm
(a tin lid is suitable)
battery
bell
wiring
nail (stylus)
plastic insulating tape
mirror
screen (wood or card)
clamp and stand
sharp knife

a experimental subject (E): 1 trial using left hand (1st trial)
b control subject (C): 1 trial using left hand (1st trial)
c E: 50 practice trials using right hand (while C rests)
d E: 1 trial using left hand (2nd trial)
e C: 1 trial using left hand (2nd trial)

This is an example of how each subject's results should be recorded.

Subject: Experimental/Control (delete as appropriate)

Trial	*Time* (*secs*)	*Errors*
1st	60	16
2nd	50	10
Difference	+10	+6

Note that a positive difference is an improvement and vice versa. The results are collated to show the total improvements of the experimental group and the control group. The difference between the improvements of the experimental and control groups will indicate the improvement in the left hand which is due to practice with the right hand.

1 Why is a control group necessary in this experiment?

2 Can you think of any everyday skills where the results of training might be transferred from one hand to the other?

Habituation is a type of learning. It occurs when an animal ceases to respond to a stimulus which is being repeated because it has learnt not to react.

Place an earthworm inside a piece of glass tubing about the same length as itself. Lay the tubing alongside a ruler. Touch the head of the worm and measure the distance which it withdraws. Repeat the procedure each time the worm relaxes.

1 Does the earthworm become habituated, or in other words, does its response become reduced or even cease?

You will need

glass tubing
ruler

2 Repeat the experiment with other worms. Do they all behave in the same way?

3 What do you think is the value of habituation?

A person's memory span is the number of individual items which he can reproduce correctly and in order immediately after hearing or seeing them once. In the following memory span test, the items are digits.

4759
28574
385793
7598385
18508749
481946390
1756385739
95716883634
958165083497

Notice that each series contains one more digit than the series before it.

You can use this test in the following way.

1 Read out the first series of digits to the subject (once only).

2 Ask the subject to write the series down in the correct order.

3 Check the result.

4 Repeat this procedure for each series.

The subject's memory span is the greatest number of digits which he was able to recall in the correct order. Do not ignore the result of a series because he did not finish the series.

The results of one test may not be reliable, so you should make up some more tests, re-examine the subject and then find the average of his results.

If this test is carried out in a school, then a whole class can be tested at the same time to find the range and average memory span.

If possible, it would be interesting to find out whether memory spans increase or decrease with age. For this you will need to test a number of classes of different ages.

You will need

pencil
paper

9 Problem solving

Humans can solve problems better than any other animal. This is because of our ability to look ahead and see the results of actions before we make them. Our predictions are expressed symbolically in images or words.

Try to see how this is true by solving the maze problem. Find a way from the entrance to the exit without entering a cul-de-sac.

Another method of solving maze problems is by trial and error. Obtain a piece of paper which is at least

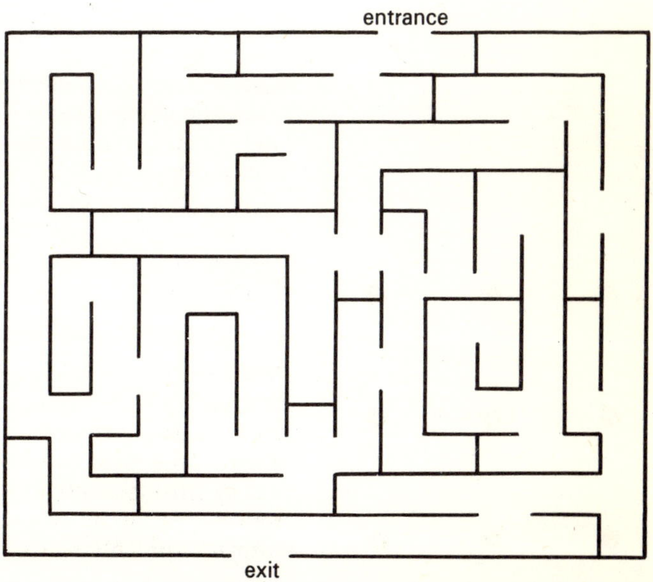

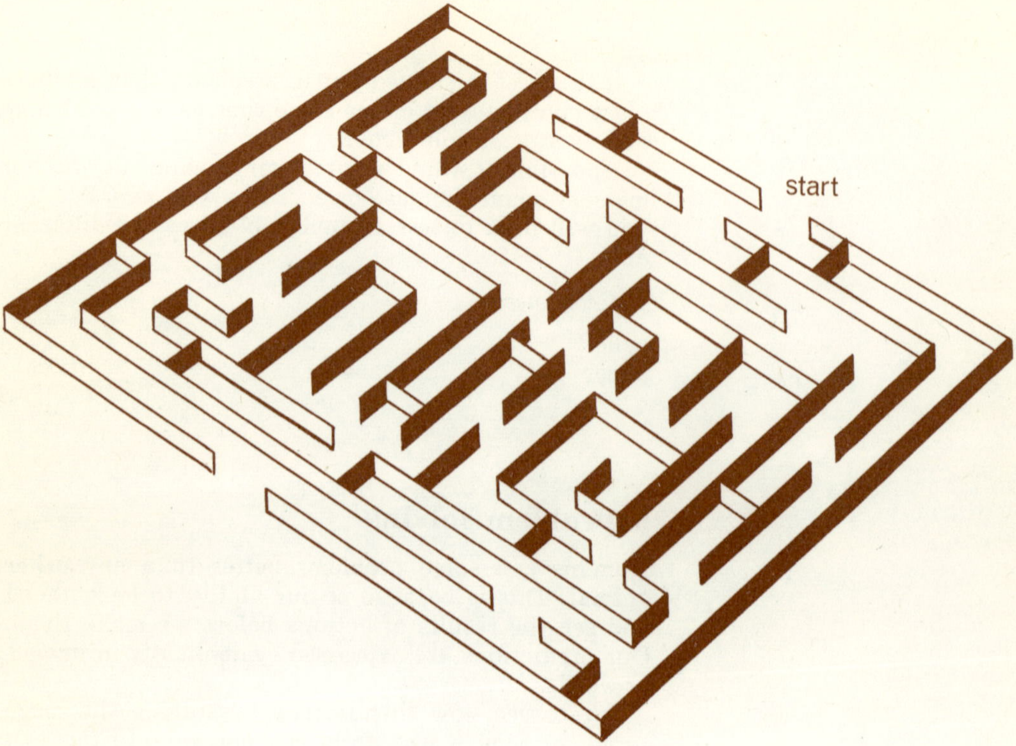

twice as big as the maze shown above. Cut a hole with a diameter of about 7 mm in the centre of the paper. Place the paper over the maze so that the 'start' can be seen through the hole. Now try to find your way through the maze by moving the hole along the lines of the maze. When you come to a cul-de-sac, do not uncover the maze, but trace your way back and try another line. After making many errors you will come to the end of the maze—but do you remember the route? Test your learning by trying to work through the maze again.

Compare the difficulty of trying to solve a problem by the method of trial and error and the first method where you were able to look ahead and anticipate the results of your action.

Problem solving becomes difficult when there is some kind of 'block' which prevents us seeing the solution. Look at the problem with nine dots arranged in a square. The problem is to connect them by drawing four continuous straight lines without lifting pencil from paper. Most people find this quite difficult. The solution and the mental 'block' involved are on p. 144.

10 Breeding behaviour of the stickleback

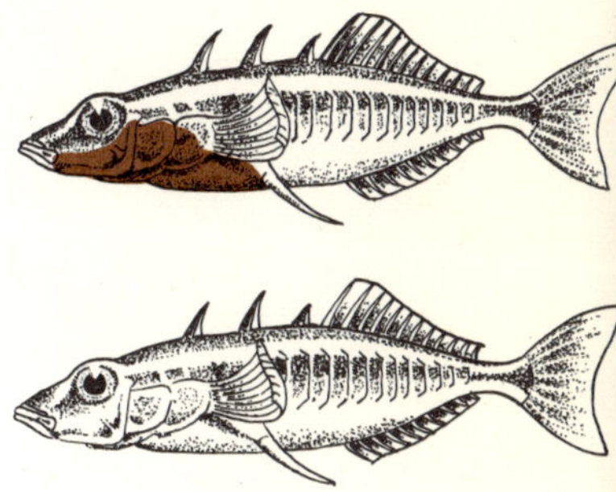

Sticklebacks are useful animals for studying behaviour. They are easy to keep in the laboratory—you can find many books which will tell you how to set up an aquarium and how to look after sticklebacks. The stickleback's behaviour is very interesting during the spring when they begin breeding. Keep some of these fish and try to make the following observations.

1 In early spring reproductive hormones cause changes in the appearance of the male stickleback. What colour do the eyes become? What colour does the belly become?

2 The male builds a nest to protect the eggs which will be laid later. How does he build a nest and what materials does he use?

3 Put two males in a tank and separate them by dividing the tank into two using an opaque glass partition. Each will establish his territory (breeding area) in his own half of the tank. Remove the partition after the fish have had time to settle down. What does the stickleback do when the other male trespasses on its territory? Why does the male establish and defend a territory? Why is it more useful to the stickleback species if the male can defend its territory by threatening rather than fighting other males?

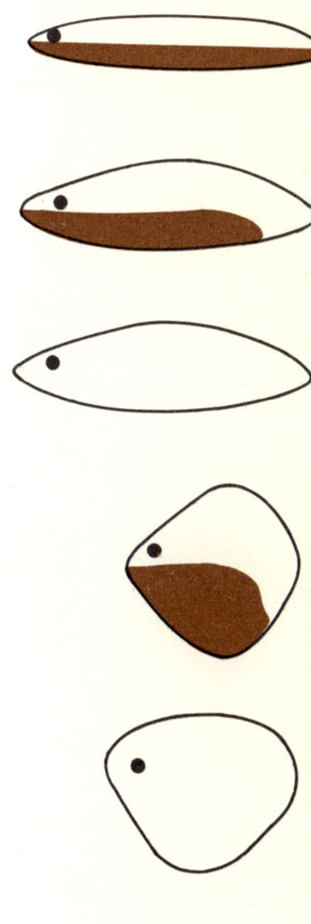

4 What is it about the other male that starts off the male's defence behaviour? Is it the intruder's shape or its colour? Make the models shown in the diagram. Present each model to the male stickleback. Which models does the male respond to? You should now be able to decide which stimulus (shape or red belly) starts the defense behaviour.

5 Keep a male and female stickleback in a tank. When the nest has been built you may observe the ritual courtship and reproductive behaviour of these fish. How does the male recognize the pregnant female? How does the female recognize the male? Notice the movements which the male makes to attract the female to the nest. Notice how the male points out the entrance to the nest which the female then enters. Once in the nest, the female is stimulated to lay eggs by certain movements of the male. What are these movements? When the female leaves the nest, the male then enters it for an essential purpose. What is this purpose?

6 For the next 7–10 days the male aerates the developing eggs by fanning water through the nest. Which of his fins does he use for this purpose?

7 When the eggs hatch, the baby fish are kept in the nest by the male. If a young fish does stray, you may be able to observe how the male returns it to the nest.

You will need

aquarium (60 × 30 × 30 cm)
coarse sand or fine gravel
aquatic plants (including algae for nest building)
glass cover (to cover aquarium)
four corks (to support cover)
opaque glass partition
models of male stickleback

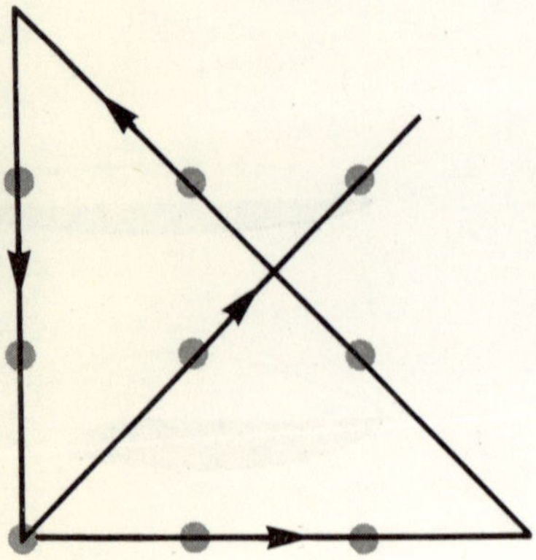

The solution to the problem on page 142
The 'block' is not realizing that the lines may be extended beyond the dots.

11 The behaviour of a small mammal

Much of our knowledge about animal behaviour has come from direct observation, both in the field and in the laboratory. Small rodents are easy to keep and their behaviour is interesting. One popular small rodent is the Mongolian Gerbil (*Meriones unguiculatus*).

Although gerbils can normally be kept in a wide variety of cages, a large aquarium would be most suitable for the purposes of this investigation. A lid for the aquarium can be made from wood and perforated zinc. Since gerbils are burrowing animals, the aquarium should be half-filled with a mixture of hay, peat and sand. If a flower pot with the end knocked out is pushed into the surface of the mixture, it will often be used as the entrance of a burrow. Gerbils may be fed on a pellet diet, supplemented with bits of apple, nuts and vegetables. Although gerbils do not drink much water, a drinking bottle should be supplied, as should a piece of wood for gnawing.

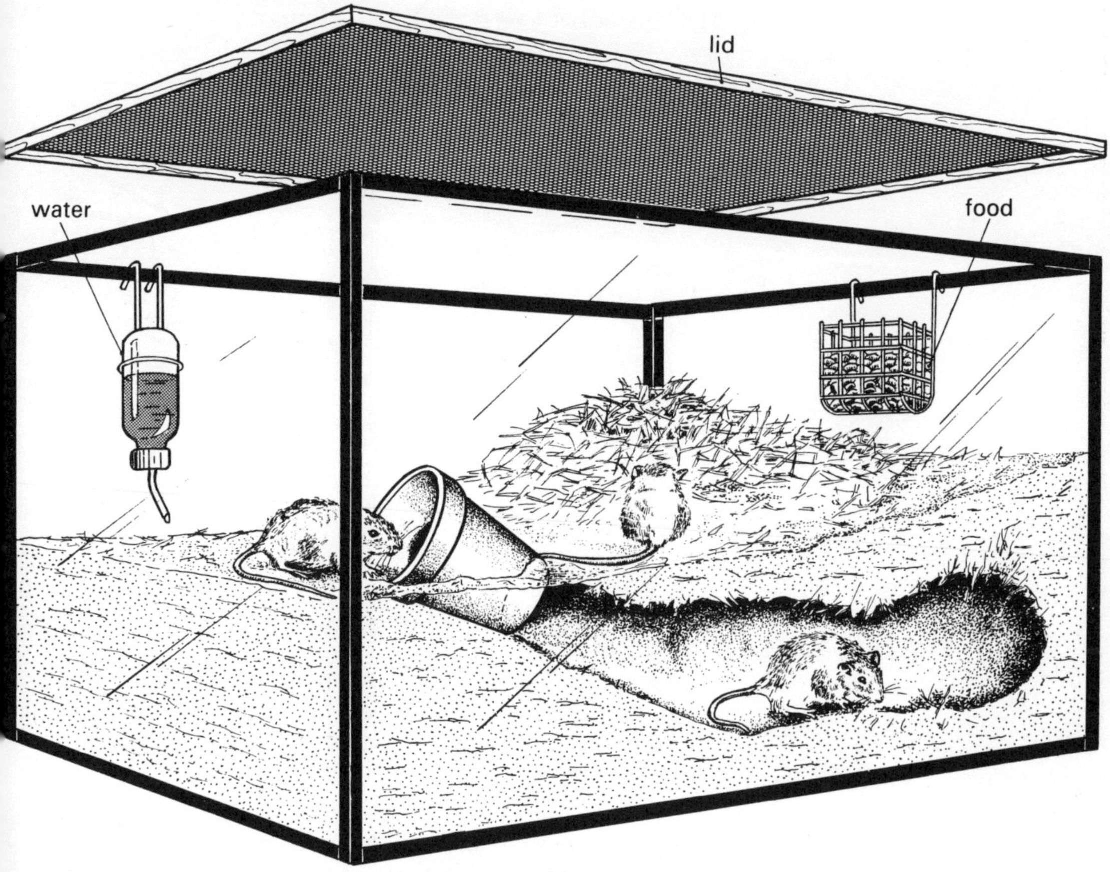

1 When you feed the gerbils, you may notice that they seem to prefer certain foods. Can you design an experiment to confirm or dispel your suspicion? When a gerbil is eating, note the body position and the use it makes of its front limbs. Do gerbils hoard their food?

2 Are gerbils inquisitive animals? Try putting a gerbil in a large box with strange objects, such as a wheel, a see-saw, newspaper, a cardboard box with a hole in it, and observe its behaviour. Is your gerbil attracted to movements and noises? Try waving a piece of paper and rattling a bunch of keys.

3 Your gerbil will probably spend much of the day asleep. Does this mean that it is more active at night? An actograph was used to study the patterns of activity of a variety of invertebrates. This apparatus can be adapted to enable you to study the pattern of activity of a gerbil. Put the gerbil in a plastic cage set up as

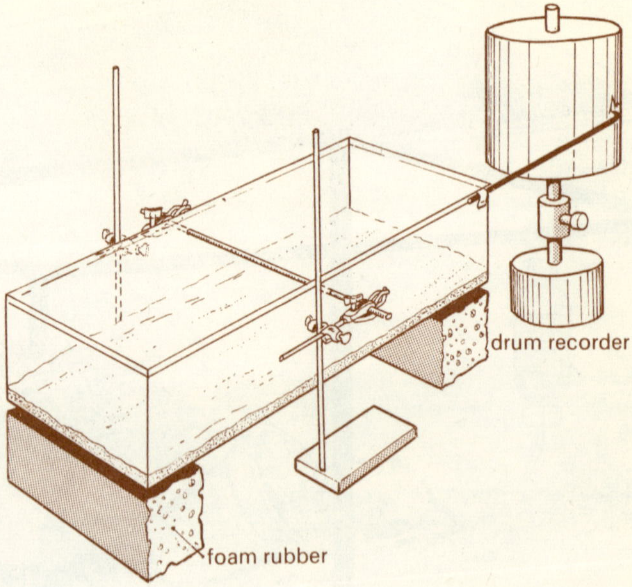

shown in the diagram. The gap between the ends of the cage and the foam rubber pads should not exceed one centimetre. Use the recording drum as before and this should help you to answer the question.

4 When a gerbil is burrowing, notice how the legs are used to move the material. Notice also where and how the gerbils build their nests. Are both sexes involved?

5 Gerbils are sociable animals, and therefore you should not keep one on its own. If you keep a male and female gerbil together, they should eventually breed. How do gerbils care for their young? You should also notice the appearance and behaviour of the young gerbils and study their rate of development.

We cannot cover every aspect of behaviour here and in the course of your study, you will discover many things for yourselves.

You will need

large aquarium	plastic cage
wood	foam rubber
perforated zinc	wire
hay	kymograph and accessories
peat	flower pot
sand	drinking bottle

large box containing objects such as wheel, newspaper

12 The social behaviour of ants

The study of social behaviour in mammals is difficult to undertake in the laboratory. For this reason, ants have been chosen as the subject for investigations into social behaviour.

First it will be necessary to set up a formicarium. Obtain two sheets of glass approximately 30 cm square. Glue strips of wood about 4 mm thick and 10 mm wide round the edges of one sheet of glass, leaving a gap of 10–15 mm in one side. Sieve some soil and add sufficient water to make it moist, but not saturated (sterilized soil is best). Pour some of the soil onto the prepared glass plate and level it off by running a ruler over the top of the wooden surround. Sufficient soil should be used to cover about two-thirds of the glass, leaving clear the area adjacent to the hole in the surround. Place the second sheet of glass over the top and secure with *passe-partout* tape or two strong elastic bands.

Prepare a base for the formicarium. Obtain a sheet of glass or metal 30 cm by 60 cm. Stand this on wooden blocks on a leak-proof tray at least 40 by 70 cm. Pour a shallow layer of engine oil into the tray.

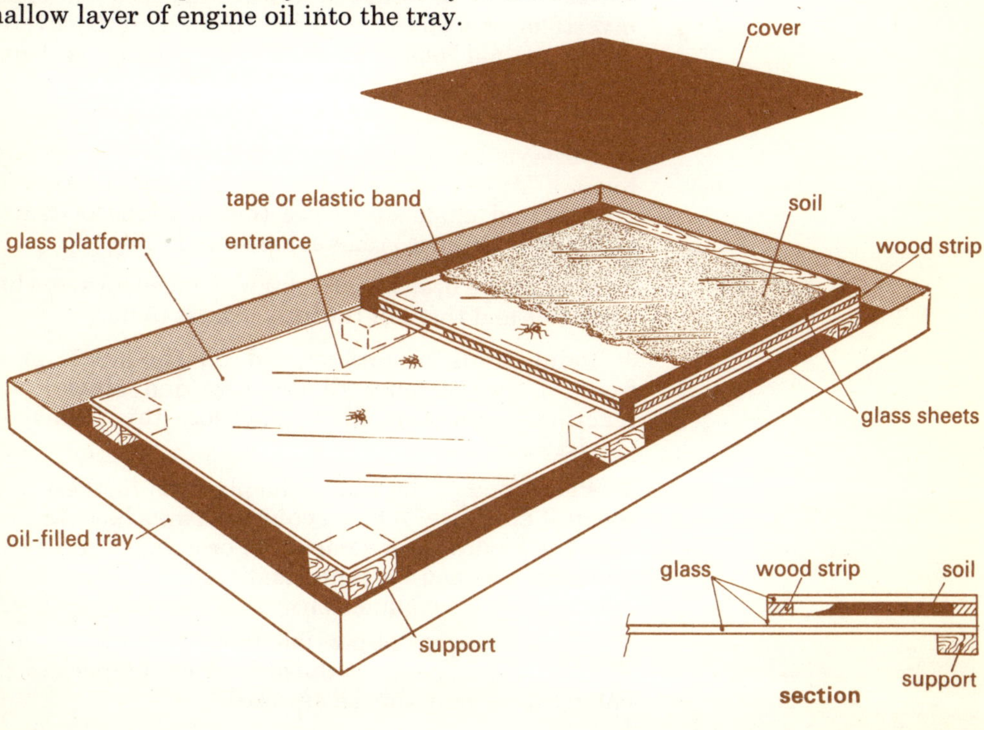

Collect some ants. Almost any species will be satisfactory but you will most likely find a community of a species which raises a mound above its nest. Such an ant is the yellow ant which has a light brown head and body and yellowish legs. It is very common in fields and meadows.

Dig out a nest of ants and place it in the middle of a sheet of white paper, you will be able to sort through the soil more easily (cotton wool tucked round wrists and ankles will protect the collector). Find, if you can, one large-bodied queen, and collect fifty or sixty workers and similar numbers of larvae and pupae. Drop these through the gap into the formicarium (this requires patience!) and seal the gap with a plug of cotton wool. Cover both sides of the formicarium with black paper and allow to stand overnight. Then lay the formicarium on the stand you have prepared and remove the cotton wool plug.

Once the colony is established, it will be necessary only to feed the ants on spots of honey, scraps of meat, etc., and to add a few drops of water to the soil when it appears to be drying out.

The platform area in front of the formicarium entrance can be used as a working area for many investigations. The formicarium itself should be covered unless actual observation of the colony is being undertaken.

Carry out some of the following investigations.

1 Observe the day-to-day activities of the colony. Keep a record of progress in the construction of tunnels and chambers. Note the positions of larvae and pupae. Do these remain constant?

2 If you have been able to obtain a queen, observe her activities and those of the ants closest to her.

3 Paint trails of formic acid on the platform in front of the entrance to the nest and observe the reaction of the ants. How long does this behaviour persist?

4 Place food at the end of the platform furthest from the nest entrance. When a column of ants has begun to visit the food, wipe a damp cloth or a finger across the platform between entrance and food. Observe results and suggest an explanation.

5 Place a variety of possible foods along the edge of the platform. Try to establish food preferences of the colony. How may this be assessed?

6 Drop a dead fly onto the platform. What happens?

7 Mark individual members of the colony with *small* spots of cellulose paint or nail varnish of different colours. Observe the behaviour of these individuals. Where do they go in the nest? How often do they visit the food, etc?

8 Place a few ants from a different colony onto the platform and watch the behaviour of the established colony.

9 Remove one member of the colony and place it in a specimen tube containing cotton wool to which one drop of clove oil has been added. Return the ant to the colony and observe.

These are only a few of the possibilities for investigation, and many more ideas will occur to you as you proceed.

You will need

two glass sheets 30 × 30 cm
strips of wood 4 × 10 mm
ruler
elastic bands
sheet of metal or glass 30 × 60 cm
wooden blocks
leak-proof tray 40 × 70 cm
engine oil
white paper
cotton wool
black paper
formic acid
cellulose paint
specimen tube
clove oil and soil

13 A problem of adolescence

A The background

John Smith is fifteen years old and lives in a comfortable home in a pleasant street. He goes to the local comprehensive school and his work and behaviour have always been good. John is an only son and his parents have always made an effort to bring him up in the manner which they think proper. Mr Smith is a bank manager and Mrs Smith used to be a teacher.

B A conflict arises

Over a period of time, John develops attitudes which differ from those of his parents.

John	Parents
1 School is a waste of time.	A good education is never wasted.
2 I don't want to wear school uniform.	School uniform is practical.
3 Pocket money is mine to spend as I wish.	We know best about money.
4 I am old enough to stay out late.	You must be in by 10 pm.
5 Jane is a great girl.	She is not suitable for you.
6 I'd like a part-time job.	You have not got time.

7 I'd like to go to a pop festival.	We don't want you going to that sort of place.

C A crisis occurs

Without telling his parents, John goes to a pop festival with some friends. Mr Smith receives a telephone call from the police. John has been found in possession of some cannabis and is being held at the police station. Mr Smith arrives at the police station and the following conversation takes place.

Father 'Is this true?'
Son 'Yes.'
Father 'Why did you do this?'
Son 'I don't understand what you mean.'
Father 'Why did you go to the festival when we told you not to?'
Son 'Because my friends were going.'
Father 'I suppose you went with Jane. We always said she was a bad influence.'
Son 'This is not her fault. I asked her to go and anyway I bought the cannabis.'
Father 'We knew you were mixing with the wrong type.'
Son 'They are not the wrong type. They are my friends and I like them. You would not like anything I want—so I don't care what you think.'
Father 'We have always done our best for you. I just don't understand you—I don't know what has gone wrong.'

What has gone wrong?

1 Looking at section B, try to think of reasons why John developed different attitudes than those of his parents.
2 Could John and his parents have modified their attitudes and reached a compromise?
3 Considering the conversation in section C, how do you think the relationship between John and his parents will develop?
4 Is disagreement between parents and their children, *a* common, *b* natural, *c* desirable?

14 Exhibition of behaviour problems

There are many social problems associated with human behaviour, of which two of the most important are mental illness and addiction to drugs. As these problems affect such a large proportion of the population, it is important that we should all learn as much about them as possible. By constructing an exhibition, you will not only learn about these problems yourself, but you will be able to pass the information on to others.

When you put on an exhibition, it is essential to gather accurate information and present it in an interesting way. The form of the exhibition will depend on the time and facilities which you have available. For instance, the exhibition may include pictures and writing, but if you have more facilities, it could include films, slides, tape recordings, visiting lecturers, etc.

In setting up your exhibition, you should bear the following topics and activities in mind.

1 *Mental illness, including the mentally handicapped*

How many people are mentally ill?
Is this an increasing problem?
What causes mental illness?
What are the different types of mental illness?
What are the symptoms?
Why is mental illness a *social* problem? How does it affect the family? How does it affect society?
How is mental illness treated?

Can you interview a local doctor? Try to find out what proportion of his patients are suffering from slight mental illness. What medicines are available? Where would the doctor send people who, in his

opinion, were in need of special help? Are there any mental hospitals or psychiatric units in your area? If so, where are they? Would someone from one of these hospitals be prepared to visit your school? If not, perhaps your local educational psychologist could give you a talk, which you could record for later use.

Your exhibition should include some information on what the individual can do to help the mentally ill and handicapped. You could find out about local and national charities and try to discover if there is anywhere in your own area where you could go to help. The Medical Officer of Health may be able to help. He may also be prepared to talk to you about how you can avoid mental illness.

Some of you may like to present a historical display to show changes in the attitude to, and treatment of, mental illness.

One child in every eight will need mental hospital treatment at some time during his or her life.

At the end of this book you will find some useful addresses to which you can write for extra information.

2 *Drug addiction*

Alcoholism

How many alcoholics are there in Britain?
Is alcoholism a growing problem?
What are the effects of alcohol on the body and behaviour of a person?
What is the consumption of alcohol in this country?
Why is drinking such a popular social pastime? What turns a social drinker into an alcoholic? What are the attitudes of society to alcoholism?
How do alcoholics affect the lives of their families and society?
Perhaps you have seen meths drinkers. Who are they? Where do they come from? How do they live? Why do they often gather together?
What treatment is available for alcoholics? Try to find out something about the work of Alcoholics Anonymous.

Addiction to other drugs

You may like to apply the above questions to guide you in investigations into addiction to other drugs.
In addition, you should find out what drugs are used, where they come from and how they are taken.
Which age-group is probably most vulnerable to drug addiction and why?
Can drug addicts be cured, and how?

Here are some useful reference books and addresses

Here are some useful reference books and addresses—

Keeping Animals in Schools (A Handbook for Teachers) HMSO 1970

Experiments in Animal Behaviour by Marguerite D. Hainsworth, Macmillan 1967

Handbook on the Care and Management of Laboratory Animals Part 11 (Rodents, Lagomorphs and Insectivores), UFAW (see adress below)

The University Federation of Animal Welfare (UFAW)
230 High Street, Potters Bar, Herts

National Council on Alcoholism,
45, Great Peter Street, London SW1P 3LT

Association for the Prevention of Addiction Ltd,
9 James' Street, London WC2E 9LA

Institute for the Study of Drug Dependence,
Kingsbury House, 3 Blackburn Road, London NW6 1XA

National Association for Mental Health,
22 Harley Street, London W1N 2ED

National Society for Mentally Handicapped Children,
17 Pembridge Square, London W2 EP

Acknowledgements

Page 4 S. Csordas; 6 Geoffrey Bridget; 8 H. Newman, photo S. Dalton; 13 G. M. Villermet; 14 S. Charlton; 15 Mary Evans Picture Library; 16–17 G. Neuweiler; 19 *World Medicine*; 20 Metropolitan Police; 23 Wembley Stadium; 24 Barnet Saidman FRPS, FIIP; 26 (top) H. Newman, photo S. Dalton; (bottom) J. M. Mennell; 27 Alphabet and Image, photo E. Winson; 31 and 36 E. D. Lacey; 38 Melody Maker and Associated Press; 40 and 41 Alphabet and Image; 44–45 Trustees of the British Museum; 46 Alphabet and Image, photo M. Price; 48 (top) Alphabet and Image; (left) H. Newman, photo I. Polunin; (right) J. Woodward; 49 J. Woodward; 50 H. Newman, photo J. Blossom; 52–53 copyright *Woman* magazine Syndication International; 54 C. von Furer-Haimendorf; 55 South African Information Service; 58 and 59 Heather Angel; 60 S. Csordas; 61 (top) Heather Angel; (bottom) E. Blechschmidt, from *Vom Ei Zum Embryo*, Deutsche Verlagsanstalt Stuttgart 1968; 62 Queen Charlotte's Maternity Hospital; 63 H. Newman, photo W. Cairns; 65 FAO; 66 Heather Angel; 68 Alphabet and Image, photo A. Ginet; 70 *World Medicine*, photo K. Morris; 71 Alphabet and Image; 73 USIS; 74 *The Times*; 75 Alphabet and Image, photo A. Ginet; 77 Keystone Press; 79 S. Poett; 82 Iyengar Yoga Institute; 84 Roman Picture Library; 85 WHO; 87 Keystone Press; 88 Bow Mission; 89 J. Cavanagh, Chelsea College of Art; 90 and 91 National Association of Mental Health, photo T. Othen; 92 J. Mennell; 93 and 94 (right) National Association of Mental Health; 94 (left) *World Medicine*; 96 Heather Angel; 99 H. Newman, photo S. Dalton; 100 and 101 (left) C. Butler; 101 (right) H. Newman, photo S. Dalton; 102 C. Butler; 105 S. Csordas; 107 Alphabet and Image, photo M. Price; 112 Alphabet and Image; 113 Trustees of the British Museum; 115 S. Charlton; 118 National Association of Mental Health, photo J. Brooke; 132 J. Mennell and G. Robinson; 150 *Melody Maker*; 152 National Association of Mental Health.

Index